The Mathematical Works of Leon Battista Alberti

Translations and Commentary by
Kim Williams, Lionel March and **Stephen R. Wassell**

Foreword by
Robert Tavernor

With contributions by
Richard Schofield and **Angela Pintore**

 Birkhäuser

Editors
Arch. Kim Williams
Kim Williams Books
Corso Regina Margherita, 72
10153 Turin (Torino)
Italy
kwb@kimwilliamsbooks.com

Prof. Lionel March
The Martin Centre
University of Cambridge
1-5 Scroope Terrace
Cambridge CB2 1PX   UK
lmarch@ucla.edu

Prof. Stephen R. Wassell
Dept. of Mathematical
Sciences
Sweet Briar College
Sweet Briar, VA 24595 USA
wassell@sbc.edu

Mathematics Subject Classification MSC2010: 01A40, 51-03

ISBN 978-3-0346-0473-4   e-ISBN 978-3-0346-0474-1
DOI 10.1007/978-3-0346-0474-1

Library of Congress Control Number: 2010935727

Cover design: deblik

Cover image based on Leon Battista Alberti's self-portrait plaque in bronze, *c.* 1435, now in the National Gallery of Art, Washington, D.C.

Printed on acid-free paper

Springer Basel AG is part of Springer Science+Business Media

www.birkhauser-science.com

Contents

Foreword
Robert Tavernor vii

Introduction
Kim Williams, Lionel March and Stephen R. Wassell 1

Leon Battista Alberti, *Ex ludis rerum mathematicarum*

    Transcription Codice Galileiana 10 fols. 1r–16r 10

    English translation of *Ex ludis rerum mathematicarum* 11

    Transcriber's note
    Angela Pintore 71

    Translator's note
    Kim Williams 72

    Commentary on *Ex ludis rerum mathematicarum*
    Stephen R. Wassell 75

Leon Battista Alberti, *Elementi di pittura*

    Elements of Painting
    Kim Williams and *Richard Schofield*, trans. 142

    Commentary on *Elements of Painting*
    Stephen R. Wassell 153

Leon Battista Alberti, *De componendis cifris*

    On Writing in Ciphers
    Kim Williams, trans. 171

    Commentary on *On Writing in Ciphers*
    Lionel March 189

    Notes on the Translation of *On Writing in Ciphers*
    Kim Williams 200

Leon Battista Alberti, *De lunularum quadratura*

    On Squaring the Lune
    Kim Williams, trans. 203

    Commentary on *On Squaring the Lune*
    Lionel March 209

Bibliography 215

About the authors 223

Foreword

Robert Tavernor

Leon Battista Alberti (1404–72) was an intellectual with a deep knowledge of many diverse subjects, which his range of writings record. His *Ex ludis rerum mathematicarum*, which this new English edition translates from the Italian, is critical to the profound understanding of mathematics and geometry that Alberti had acquired during the 1430s and 50s, and which underpinned his rigorous approach to the visual arts: painting, sculpture, urban design and – most tangibly (since fine examples of his buildings still survive) – his architecture, which influenced leading architects of the Italian Renaissance, and beyond.[1]

Born illegitimately in Genoa, Italy, into an exiled Florentine banking family, Alberti had an insecure and peripatetic start to life, spending his formative years in Padua (where he boarded in the Gymnasium led by the exceptional classical scholar Gasparino Barzizza), and Bologna (where he studied canon and civil law at the city's great university). Pope Martin V lifted the ban on the Albertis' exile in 1428, the year of his graduation, when it is generally assumed Alberti first entered Florence as secretary to Cardinal Albergati, Bishop of Bologna. But he was in Rome that year as secretary to Bishop Biagio Molin, head of the papal chancery, who nominated Alberti to the College of Abbreviators in the Papal Court (*abbreviatori apostolici*). This was evidently an environment that suited him, as he remained there drafting papal briefs for the next thirty-two years, which left him time to connect with the leading minds of his era, to think, write and make. The independence this way of life offered was made financially secure when he was appointed rector in 1448 of the small parish of San Lorenzo in Mugello, outside Florence.

His writings were literary and technical in character. His first literary triumph was a comedy, *Philodoxeos* (Greek for 'lover of glory', 1424), written at the age of twenty and – worked in the antique manner – he succeeded in passing it off for a long time as the work of a fictional Latin author, Lepidus – meaning 'the joker'. The *Intercenales* ('Dinner Pieces', 1429), a light hearted collection of dialogues and fables, was followed by a dissertation on the life of the intellectual, *De commodis atque litterarum incommodes* ('The Pleasure and Pain of Letters', 1429–30). Its sobriety was balanced by a discourse on the character of love, *Amator*, and *Ecatonfilea* ('A Hundred Loves'), written at about the

---

[1] This preface is based on material set out in *On Alberti and the Art of Building* [Tavernor 1998].

same time, and *Deiphira* (ca.1429–34?), which explores ways of escaping failed relationships. This fascination with human nature culminated at this time with the Italian work, *Della famiglia* (1433–4), which explored family relationships and the value of virtue (*virtù*) – educated reason as an antidote to mere opinion – which he adopted as a guiding principle.

Alberti's search for truth and the certitude of reason led to his careful work on Law (*De iure*, 1437) and proposed reforms to the Tuscan dialect (in *La prima grammatical della lingua Toscana*, around 1450) by introducing latinised terms and phrases, as well as to scrutinise the mathematical and geometrical principles on which the greatest human artefacts are based. Extending these principles into the visual arts, he codified the early technical development of perspective by the Florentine circle of artists and architects, led by Filippo Brunelleschi in the early part of the fifteenth century, in *De pictura* (On Painting, 1435). Alberti dedicated the Italian translation of this work (*Della pittura*) to Brunelleschi in 1436, the year that the hugely impressive dome of Florence Cathedral was consecrated by Pope Eugenius IV. The mathematical and geometrical prowess that Brunelleschi displayed through this dome was innovative as well as visually remarkable. Alberti wrote in his dedication:

> [was it not] vast enough to cover the entire Tuscan population with its shadow, and done without the aid of beams or elaborate wooden supports? Surely a feat of engineering, if I am not mistaken, that people did not believe possible these days and was probably equally unknown and unimaginable among the ancients.[2]

Alberti's treatise on painting was highly original too, as he was the first to write coherently about a theory underlying the art of painting, as opposed to providing a history of painting (which earlier classical authors had done, such as Pliny the Elder in his *Natural History*). Central to his theory was the construction of three-dimensional space on a two-dimensional plane using monocular, one-point perspective. Conceived as visual rays – lines that appeared to emanate from the artist's eye in the form of a pyramid to a measured ground plane constructed of orthogonals that angle towards a point of perspective or vanishing point – and viewed through a picture plane, objects beyond appear in scale relative to one another and the viewer. His companion volume for the visual arts was on sculpture, *De statua*, which was probably written at about that time or soon after and focuses on the techniques for measuring and proportioning the three dimensions of human figures accurately (real or as existing sculptures) and in relation to a notion of idealised human form.[3] He invented a measuring tool for this purpose – the *finitorium* – composed of a measuring disc, or 'horizon', measuring 3 notional 'feet' in diameter, a moveable calibrated radius (*radium*) pinned to its centre, a plumb line, and an *exempeda* or measuring rod. The *finitorium* was placed above the vertical axis of the subject being measured and the various joints of the limbs relayed back through the *exempeda*, plumb-

---

[2] [Alberti 2004a: 35].
[3] See [Tavernor 2007: 31–32].

line and radius in order to record three-dimensional points in space in relation to the disc.[4] Alberti translated this theory into practice himself, and he painted and sculpted; though little of his work survives by which to judge him, except for a small, finely wrought bronze self-portrait plaque of his head in profile and modelled in relief.[5]

His search for reason also led to him to create a new tool with which to make the first accurate physical survey of Rome, recorded in *Descriptio urbis Romae* (ca.1444), which developed from the *finitorium* of *De statua*. It was large in size: the sculptor's measuring disc had a diameter of 3 feet and a circumference divided into 6 degrees and sixty minutes, while for the surveyor he devised a horizon with a diameter of 10 feet and a circumference divided into 48 degrees and 192 minutes. The disc was taken to a vantage point and Alberti used the Campidoglio in Rome (and perhaps the summit of the tower on the Palazzo Senatorio) for a clear view. He constructed his map by rotating the *radium* at its centre to topographical features – including churches and ancient sites, the city walls with its towers, and the course of the River Tiber – and then measuring distances by pacing and through geometry. He later applied a similar idea of a calibrated disc for a quite different purpose, as a cipher wheel for writing and decoding cryptic messages. It is described in *De componendis cifris* (1465, and translated into English here) as composed of a larger wheel with a smaller one inset and sharing the same centre point.

Taken as a whole, Alberti's literary output reads as an orderly progression through the intellectual and social world he inhabited, with each work building on the preceding and with linked works on related subject areas. Thus, ideas expressed in a Latin pamphlet *De punctis et lineis apud pictores* (circa 1435) were developed more fully the following year in his *Elementi di pittura* (translated into English here); which was when *De pictura* was being translated into Tuscan for Brunelleschi. He also wrote a separate work on geometry, *De lunularum quadratura* (ca. 1450) around the time he was preparing his *Ludi rerum mathematicarum*. His mastery of these subjects – mathematical principles and their three-dimensional consequences through geometry – led to his involvement in major physical projects, including excavating monuments and even an attempt at raising the remains of two ancient Roman ceremonial barges from the floor of Lake Nemi in the Alban Hills above Rome.[6]

His most profound literary and technical synthesis was his lengthy treatise on architecture, *De re aedificatoria*. It was probably begun around 1440 as a literary commentary on Vitruvius's *De architectura*,[7] but it evolved into a full blown architectural treatise of considerable significance in its own right over the next decade or so, as he built major buildings in Florence, Rimini and Mantua – his church designs in particular influencing generations to come, including Andrea Palladio. They represent an explicit expression of his concern for number, measure and proportion set within a specific

---

[4] For an exploration of the translation of theory into practice of *De pictura* and *De statua* in the perspective of Piero della Francesca's painting, *The Flagellation*, see R. Tavernor, 'Contemplating Perfection through Piero's Eyes' [Dodds and Tavernor 2002: 78–93].

[5] [Tavernor 1998: 31–34].

[6] [Tavernor 1998: 15].

[7] See *On Architecture*, R. Schofield (transl.) with an introduction by R. Tavernor [Vitruvius 2009].

contemporary cultural context – visual as well as intellectual.[8] His approach to architecture would not have been the same without his mastery of mathematics and geometry.

Produced before the age of printing, all Alberti's writings appeared in manuscript form, which were copied by scribes and often dedicated by him to selected patrons, princes and friends. Consequently, there were very few in circulation as they were expensive to reproduce. Their limited circulation was extended by owners of the manuscripts exchanging them for others, or through public readings: Alberti's treatise on architecture was almost certainly meant to be read out aloud to an assembled group.[9] Frequent copying led to the corruption of the original through errors, misinterpretation or deliberate 'improvements' by scribes, and diagrams posed a particular challenge for professional copyists more familiar with text. Authors of technical manuscripts were understandably wary. Alberti expressed this concern in his treatise on architecture, which – like all the manuscripts he authored in his lifetime – was unillustrated. Referring to the design of the classical column, which he described only with words and numbered proportions, he remained anxious that even the numbers would be copied incorrectly if abbreviated with Roman numerals, a concern he conveyed to his potential scribes:

> Here I ask those who copy out this work of ours not to use numerals to record numbers, but to write their names in full; for example, twelve, twenty, forty, and so on, rather than XII, XX, and XL.[10]

Faithfully, even the first printed edition of Alberti's *De re aedificatoria* published posthumously in 1486 followed this request and was left unillustrated.[11] Technology has of course advanced considerably since then, and we trust that the precision and accuracy of modern publishing combined with the careful scholarship of the illustrated editions that follow here would have satisfied even Alberti's attention to detail.

---

[8] [Tavernor 1998: 39–48].
[9] See [Alberti 1988: xxi].
[10] [Alberti 1988: 200–201 (VII, 6)].
[11] [Tavernor 1998: 16].

Introduction

Kim Williams
Lionel March
Stephen R. Wassell

Alberti's mathematical works

Leon Battista Alberti (1404–1472) was a prolific fifteenth century polymath, a true Renaissance man. His contributions to architecture and the visual arts are well known and available in good English editions; much of his literary and social writings are also available in English – including the recent 2003 edition of *Momus* in Harvard's I Tatti Renaissance Library. However, Alberti's mathematical works are not well represented in readily available English editions.

The principal aim of this present volume is to provide accessible and understandable translations of four of Alberti's mathematical treatises for modern readers who may not be versed in Italian, and to set them in close proximity to each other in order to shed light on his particular mathematical way of thinking. We feel that a single, comprehensive volume will be extremely valuable to round out the portrait of his multifaceted intellect. In similar fashion, a second aim, through the commentaries, is to provide a way for English-speaking readers to benefit from scholarly studies in other languages, in addition to our own careful analyses. Both *Ludi matematici* and *Elements of Painting* have received a substantial amount of attention in other languages, most notably Italian and French, but they have not in English. One function of our commentary is to make available in English some of the excellent work that has been done in Alberti research years ago by scholars such as Cecil Grayson, Luigi Vagnetti and Alessandro Gambuti, and more recently by Francesco Furlan, Pierre Souffrin and Dominique Raynaud. We are very fortunate to be able to build on the considerable scholarship that has been developed over the last decade or so, spurred in part certainly by the ongoing publication of *Albertiana*, which began in 1998. Some of this scholarship appears in English as well, such as the work of Anthony Grafton on Alberti; we have benefitted from new offerings, such as Richard Schofield's 2009 translation of Vitruvius's *Ten Books*, as well as works in the larger field of history of the sciences, such as Kirsti Andersen's 2007 tome on perspective and Barnabas Hughes's 2008 edition of Fibonacci's *De practica geometrie*.

Alberti, best known to many for his architectural work, both theoretical and executed, was first and foremost a scholar. In his 'Life of Alberti', Giorgio Vasari writes: 'It is not surprising that Alberti was better known for his writings than for his works [because] he aimed to observe the world and measure antiquities, but even more so because he was

K. Williams et al. (eds.), *The Mathematical Works of Leon Battista Alberti*,
DOI 10.1007/978-3-0346-0474-1_1, © Springer Basel AG 2010

more inclined to writing than to applied works'.[1] Vasari says further, 'In painting Leonbattista produced no great works, nor of great beauty ... the existing works by his hand, which are very few, are lacking in perfection, but this not surprising, because he tended more to study than to drawing'.[2] This may depict Alberti as a dilettante, but in reality he was a polymath whose vast knowledge ranged over a number of disciplines.

Mathematics held particular interest for Alberti. As historian Paul Lawrence Rose stated in his authoritative text: 'Apart from Piero della Francesca, no artist viewed life more exhaustively in mathematical terms than Leon Battista Alberti, for whom mathematics supplied a certainty and contentment lacking in the arbitrary vexatiousness of human affairs' [Rose 1975: 6]. A century after his death, Alberti was cited by Filippo Pigafetta as one of the 'restorers' of classical mathematics in the Italian Renaissance, although he had not, like his contemporary, Piero della Francesca (1415 ca.–1492), left original works of mathematics, or engaged in publishing translations of ancient Greek and Latin mathematical texts. However, Alberti had been a close friend of the renowned mathematician and astronomer Paulo Toscanelli, to whom he had dedicated his first book of *Intercenales* (*Dinner Pieces*), which 'provide a way of relieving the mind's maladies ... through laughter and hilarity' [Alberti 1987: 15]. He had been an acquaintance of a leading 'restorer', Cardinal Bessarion, who was an advocate for translating Greek mathematics. He is also known to have moved in the circles of the philosopher Cusanus, who dabbled in mathematical speculation such as the quadrature of the circle.

The greatest astronomer of the age, Regiomontanus, reported in 1464 of hearing Alberti and Toscanelli speak concerning astronomical observations they had made together. Alberti was intrigued by mathematical instruments. These were to ornament his ideal library (*De re aedificatoria*, VIII, 9 [Alberti 1988: 287]). Alberti, in his book *De statua*, describes and discusses the adaptation of the astrolabe for use as an instrument for measuring the human figure, and in his work on surveying Rome he modifies the apparatus again for terrestrial purposes (*Descriptio urbis Romae*).

Alberti's *De re aedificatoria* was unique in its genre, the first treatise on architecture written since Vitruvius's *Ten Books*, and was thus the progenitor of all architectural treatises that followed. His treatises *Elements of Painting* and *On Writing in Ciphers* were also new contributions, but *Ludi matematici* and *On Squaring the Lune* were part of a tradition that had been on-going since antiquity. Alberti inserts himself into this tradition by mentioning earlier mathematicians with whose work he was familiar, referring to them both directly (Euclid, Archimedes, Nicomachus, Savosarda, Fibonacci) and indirectly (the 'ancients').

Alberti was both late Medieval and early Renaissance. Architecturally, he was on the leading edge of the Renaissance. Textually, he could not escape the Middle Ages. Printing had literally just been invented, but it was not yet established:

---

[1] *Non è maraviglia dunque, se più, che per l'opere manuali è conosciuto per le scritture il famoso Leon Batista ... attese non solo a cercare il mondo, et misurare le antichità; ma ancora, essendo a ciò assai inclinato, molto più allo scrivere, che all'operare* [Mancini 1917: 23–24].

[2] *Nella pittura non fece Leonbattista opere grandi, né molto belle, con ciò sia, che quelle, che si veggiono di sua mano, che sono pochissime, non hanno molta perfezzione, né è gran fatto, perché egli attese più a gli studi, che al disegno ...* [Mancini 1917: 39].

Woodblocks were almost certainly being printed on paper by the late fourteenth century, and given the relative simplicity of the technique, within two to three generations the practice became common throughout much of Europe. Intaglio printing from incised metal plates followed soon after, apparently undertaken in a serious way during the 1430s, somewhere along the Upper Rhine region of southwestern Germany [Laundau and Parshall 1994: vi].

Alberti knew of these developments, and was excited about them:

> While I was with Dati in the papal gardens in the Vatican, discussing, as usual, literary matters, we happened to praise in the most enthusiastic terms the German inventor who has recently made it possible, by means of a system of moveable type, to reproduce from a single exemplar more than two hundred volumes in one hundred days with the help of no more than three men [*On Writing in Ciphers*, p. 171 in the present volume].

Perhaps the greatest consequence of Alberti's inability to access this new technology is that he kept almost all of his writings illustration-free, including his architectural treatise, and even his treatises on painting (cf. [Alberti 2007: 3–17]). Even his *Descriptio urbis Romae* is not a pictorial map but rather tables upon tables of polar coordinates locating the main features of Rome (relative to the Palatine Hill), along with instructions on how to construct a device that will help the reader transfer the polar coordinates into an actual map. Of course, *Ludi matematici* is one of the exceptional works of Alberti's to be illustrated, as will be discussed further in the commentary.

There were two reasons for the dearth of illustrations in Alberti's written work. One is that he had inherent mistrust of copyists, believing that their versions of illustrations were almost certainly going to be problematic after some (small) number of reproductions. In fact, he mistrusted copyists even in terms of numerals, which are essentially images: 'Here I ask those to copy out this work of ours not to use numerals to record numbers but to write their names in full; for example, twelve, twenty, forty, and so on, rather than XII, XX, XL' [Alberti 1988: 200–201 (VII, vi)]. Even so, despite this injunction, the ratio eleven to four (*undecim ad quattuor*) is clearly an error for eleven to fourteen (*undecim ad quattuordecim*), as a simple comparative drawing would have shown [Alberti 1988: 219 (VII, x)]; see pp. 211–212 in the present volume for more details. The second reason is that Alberti placed a higher value on written descriptions than on images. He was not alone in this; various other authors beginning in antiquity also favoured words over images (cf., for example, [Furlan 2006: 207] and [Alberti 2007: 21]). Alberti was not of the mind that a picture is worth a thousand words (though, as will become apparent in our commentary, Alberti's *Elements of Painting* certainly could use some illustrations).

Alberti's mathematical output was not limited to the four treatises presented here. Alberti mentions another mathematical work in *De re aedificatoria:*

> How to set out angles would not be easy to explain using words alone: the method by which they are drawn is derived from mathematics and would require a graphic illustration. What is more, it would be foreign to our

undertaking, and in any case the subject has been dealt with elsewhere in our *Mathematical Commentaries*. ... [Alberti 1988: 62 (III, ii)].[3]

Some of these works are lost, such as a book on weights and measures (cf. [Grafton 2000: 82]).

Another mathematical treatise, *Descriptio Urbis Romae*, has been recently translated into English and analyzed by Mario Carpo and Francesco Furlan [Alberti 2007] in an exemplary way. The scholarly apparatus on *Descriptio Urbis Romae* compiled and constructed by Carpo and Furlan is impressive and informative, and we look forward to such offerings on other Albertian texts.

It has never been our intention to produce a critical edition. Our goals are different. Our objective is to make these treatises of Alberti accessible in English to a broad, modern, and international readership. By accessible we mean both readable and understandable, the first made possible by the translations and the second by the commentary. A further objective is to contribute to an understanding of Alberti as a true Renaissance man, as portrayed by Joan Gadol and more recently by Grafton. By making available the mathematical works of Alberti, our aim is to offer knowledge which can be inserted into the larger and growing body of Alberti scholarship.

The treatises in the present edition

The four texts dealt with here are interesting in that, while all deal with mathematics, each deals with it in a different way and with a different 'literary' approach.

Ludi matematici, the longest treatise included here (and the one which first sparked our interest in the project) is what we today would call a work of divulgation or popular science. Though written at the behest of a single noble patron, and apparently not intended for a public readership, its purpose is to explain its subject – amusing ways to use mathematical notions to practical ends – clearly but without going into detail about whys and wherefores, that is, without proofs and demonstrations. Its tone is appropriately informal, though not familiar (Alberti was after all addressing a Duke of the powerful Este family). We debated over how to translate the title *Ludi matematici*, which literally translates as *Mathematical Games* (or *Divertissements Mathématiques* as Souffrin chose to translate it) but 'games' doesn't exactly describe what the problems contained in the treatise are. (In the end we decided not to translate the title at all, *Ludi matematici* being familiar to many as it is.) A very enticing translation was *Cool Things You Can Do with Mathematics*, because in essence this is what the treatise is about, but of course that would not have been taken very seriously by our potential readership. There were even times in writing the commentary when the word 'cool' would have certainly fit the notion that Alberti was trying to convey to Meliaduse, but we decided against using the word. The present translation of *Ludi matematici* is based on a fresh transcription by Angela Pintore of fols. 1r-16r of manuscript Galileana 10 held in the Biblioteca Nazionale Centrale in Florence, first identified by Thomas Settle [1971]. The transcription and the translation appear on facing pages for ease of comparison. We are very grateful to the BNCF for generous concession of the rights to reproduce the images from that manuscript as well, which are published here for the first time.

---

[3] Alberti's *Commentarii rerum mathematicarum* may be a lost work; see [Grayson 1960].

Elements of Painting is more formally mathematical than *Ludi matematici*, and its terse prose more difficult for the modern reader to fathom. From the time of Vitruvius, many have exhorted the artist and architect to study mathematics. Alberti himself, in *De re aedificatoria,* wrote: 'Of the arts the ones that are useful, even vital, to the architect are painting and mathematics' [Alberti 1988: 317 (IX, x)]. He goes on to say:

> Let it be enough that he has a grasp of those **elements of painting of which we have written**; that he has sufficient knowledge of mathematics for the practical and considered application of angles, numbers, and lines, such as that discussed under the topic of weights and the measurements of surfaces and bodies... [Alberti 1988: 317 (IX, x)] (emphasis added).

Elements of Painting is Alberti's outline of this study. This again is not textbook mathematics – there are no demonstrations or proofs – but the tone is more formal than the *Ludi*. *Elements of Painting* was written in both Italian and Latin, as was his larger treatise, *On Painting*. As we see in the Latin transcription of *De Statua* [Grayson 1998: 309ff], Alberti mentions both *On Painting* and *Elements of Painting* in one breath in the very first sentence. *On Painting* has already been the subject of ample analysis. The commentary given here on *Elements of Painting* not only analyzes that single treatise but makes an effort to show how the two relate to each other. One difference is that while it is necessary for *On Painting* to consider Alberti's potential sources for his theories on optics, from Euclid's *Optics* to Arabic work such as Alhazen's, for *Elements of Painting* it is Euclid's *Elements* that is by far the most important of Alberti's sources. This is why we focus on *On Painting* and Euclid's *Elements* when analyzing *Elements of Painting*. The translation of *Elements of Painting* is based on the 1973 edition published in Grayson's *Opere Volgari.*

The treatise *On Writing in Ciphers* has yet another tone, less informal than *Ludi*, less terse than Elements, less mathematical than both. But here Alberti has applied his mathematical expertise to the study of language (specifically, Latin), breaking down written language into its components – words into syllables, syllables into combinations of letters, letters into vowels and consonants – and analyzing mathematically the frequency of use of individual letters and combinations of letters, producing the first polyalphabetic system of encrypted writing. The translation is based on the critical edition by Augusto Buonafalce published in 1998.

The final treatise in this volume, *On Squaring the Lune* (actually a mini-treatise, only a page in length) is a more complicated version of the classical demonstration of squaring a lune by Hippocrates. The attribution of this text to Alberti has never been confirmed, and has recently been refuted by Raynaud [2006], but we feel it is worthy of inclusion in the present volume due to its long history of being listed along with Alberti's other mathematical works. Of the four treatises we include, this is the only exercise in pure mathematics, but unfortunately it ends with the mistaken assertion that since a particular curved figure can be squared, then it is similarly possible to square a circle. The translation is based on the text published in Girolamo Mancini's 1890 *Opera inedita et pauca separatim impressa.*

Notes on the translations

From the translator's point of view, *Ludi matematici* was the most challenging of the four treatises to work with. At first reading, its informal tone belies its sometimes very difficult content, and the desire to maintain the tone while rendering the text clear and understandable proved to be a challenge. As Umberto Eco said in *Dire quasi la stessa cosa*, translating is 'saying almost the same thing'. Of all the 'tools' Alberti uses in the *Ludi* – spears, strings, cords, plumb lines, reeds, etc. – only one was left untranslated, the *equilibra*, an easily crafted instrument used by Alberti to measure both weights and levels. However, the problem with this translation was not so much one of technical terms, as it was with *Elements of Painting*, but rather one of deciding how best to make Alberti's meaning clear. For instance, what if *misurare*, 'measure', means 'take the measure of', 'judge', or 'appraise' rather than literally 'quantify the dimensions'?

As work progressed over various drafts, it was decided that while one criterion was to make the English text clear and understandable, a second criterion was to remain as faithful to the construction of the phrases as possible. There were two reasons for this. One was because although the sentence structure was sometimes a bit awkward, there is less risk of imposing an interpretation on the original phrase, which is sometimes an unwanted side effect of tidying up the prose. We thought that this particular choice, which might require some extra effort on the part of the reader, would nevertheless be helpful to those who would be able to compare the Italian text to the English text, thanks to the parallel arrangement on facing pages.

This can be made clear by the example of the opening phrase of the very first exercise of *Ludi matematici*. The original Italian from the Galileana 10 manuscript of the Biblioteca Nazionale Centrale of Florence is:

> *Se volete col veder[e] sendo in capo d'una piaza misurare quanto sia alta quella torre quale sia a piè d[e]lla piaza fate in questo modo...*

Our chosen translation is:

> If you wish, by sight only, being at the head of a square, to measure how high that tower at the foot of the square is, do like this...

But a less literal and more modern translation might be:

> If you're standing at one end of the square, and you want to measure by sight only how high the tower at the other end of the square is, do this...

Of course, the choice of word order and phrasing is a judgment call that all translators must make. We believe that with respect to the two criteria we have set, we have made the correct choice for this volume. There was less of a risk of forcing an interpretation the other three treatises included here because by their nature they were more straightforward.

If clarity was a priority for us, we know that it was for Alberti as well. In *De re aedificatoria* he vented his frustration at the lack of clarity in the Ten Books by Vitruvius:

> What he handed down was ... not refined, and his speech such that the Latins might think that he wanted to appear a Greek, while the Greeks would think that he babbled Latin. However, his very text is evidence that he wrote neither Latin nor Greek, so that as far as we are concerned he might just as well not

have written at all, rather than write something we cannot understand [Alberti 1988: 154 (VI, i)]

But although Alberti decries the lack of clarity in Vitruvius, Alberti himself is not always very clear, at least on first reading (it is very fortunate in this regard that *Ludi* is one of the few illustrated works of Alberti). Of course we have to acknowledge that by 'Alberti' we really mean only copies upon copies, since the originals are long lost. It is tempting to chalk up our lack of understanding to the long interval of time that separates us from Alberti, but a millennium and a half separated Alberti from Vitruvius, while only a third of that has passed between Alberti and the modern reader. This underscores how different the two halves of the second millennium were in the western world. By the fifteenth century an awakening had begun, but Alberti and his peers predated the impressive developments in mathematics, science, and engineering that would not occur until at least the Renaissance had gotten more under way. The intricacies involved in determining what an author from an earlier century originally meant to say have been pointed out often enough. Sometimes the contents of the discipline being dealt with itself are called into question; as Souffrin commented, it is rather difficult to have a precise idea of what fifteenth-century mathematics were.[4] But the argument that our modern mindset prevents us from knowing what an earlier author meant is sometimes used as a convenient excuse to avoid undertaking an in-depth search for objective, not solely historical, meaning.

Three of the treatises are here translated into English in their entirety for the first time (Problem XVII of *Ludi matematici* was translated by Peter Hicks in [Furlan-Souffrin 2001]); *On Ciphers* was translated into English in 1997 by M. Zanni. In Book III of *On Painting*, Alberti expressed his pleasure in being the first to set out the rules governing painting: 'I consider it a great satisfaction to have taken the palm in this subject, as I was the first to write about this most subtle art' [Alberti 2004a: 96]; I think we are equally satisfied with 'having taken the palm' to be the first to have the privilege of bringing these works of Alberti to an English-speaking readership.

---

[4] Cf. [Souffrin 1998: 88].

Acknowledgments

The authors gratefully acknowledge the contributions and advice of: Angela Pintore, Richard Schofield, Robert Tavernor, María Celeste Delgado-Librero, Sylvie Duvernoy, Mário Júlio Teixeira Krüger, Tessa Morrison, Monica Ugaglia, Laura Garbolini, Livia Giacardi, Branko Mitrović and our editor, Karin Neidhart of Birkhäuser. For help obtaining images and authorisations to reproduce them, we especially thank Dr. Antonia Ida Fontana, Director of the Biblioteca Nazionale Centrale in Florence, Dr. Piero Scapecchi, Director of the Manuscript and Rare Books Division, BCNF, and Susanna Pelle, BNCF; Dr. Piero Lucchi of the Biblioteca del Museo Correr; and Mario di Martino of the Observatorio Astronomico di Torino. For institutional support we thank Sweet Briar College.

Images from originals manuscripts are reproduce here by permission of:
Biblioteca Nazionale Centrale di Firenze
Biblioteca Classense di Ravenna
Biblioteca del Museo Correr, Venezia
Osservatorio Astronomico di Torino (OATO)
Veneranda Biblioteca Ambrosiana, Milano

Ex ludis rerum mathematicarum

K. Williams et al. (eds.), *The Mathematical Works of Leon Battista Alberti*,
DOI 10.1007/978-3-0346-0474-1_2, © Springer Basel AG 2010

1r

Se volete col veder[e] sendo in capo d'una piaza misurare quanto sia alta quella torre quale sia a piè d[e]lla piaza fate in questo modo fichate uno[1] dardo in terra et fichatelo chegli stia a piombo fermo et poi scostatevi da questo dardo quanto pare a voi ò sei ò octo piedi et indi mirata alla cima d[e]lla torre dirizando il vostro vedere a mira p[er] il diritto d[e]l dardo et li dove il vedere vostro batte nel dardo fatevi porre un poco di cera p[er] segno et chiamasi questa cera • A et più a stato et fermeza d[e]lli vostri piedi et viso quale mirasti la cima della torre mirate giu basso il pie di decta torre et quivi simile dove al dardo batte il vedere vostro fatevi porre nel dardo un'altra cera et chiamasi questa seconda cera • B • ultimo mirate qualq[ue] luogo in detta torre noto a voi atto a poter[e] facilmente misurar[e] col vostro dardo quando vi appressate alla torre come fare forse l'arco d[e]ll'uscio ò qualche p[er]niso ò simil posto in basso et come facesti mirando la cima et mirando il pie d[e]lla torre così qui fate et ponete una terza cera nel dardo dove batte la vista vostra questa è cosa nota et chiamasi questa terza cera • C • come qui vede la positura:

Ostilius   Riccius

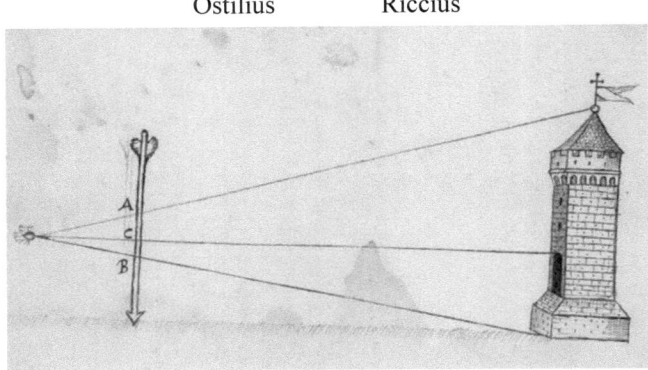

1v

Dico che quante volte enterra la parte d[e]l dardo quale stara fra la cera • B • et la cera • C • in quel altra parte del dardo quale stara fra il punto • A • et il punto • B • tante volte quella parte bassa d[e]lla torre nota a voi entra in quella di sopra ignota prima da voi et p[er] più chiareza et pratica di questa dottrina sievi questo p[er] exemplo a numeri Sia alta torre piedi cento et nella torre l'arco d[e]lla porta piedi • X trovarete nel dardo simili ragioni cioè ch[e] come quella parte d[e]lla torre dieci entra nella parte maggior[e] nove volte et inse una delle • X parte di tutta la torre cosi la parte d[e]l dardo • A • C • divisa in nove parte sara tale ch[e] la ricevera nove volte • B • C • el decimo di tutto • A • B • et cosi mai errarete purche al porre de punti voi trovatevi sempre con l'occhio el primo stato questo medesimo potete far co' un filo piombato facendo pendere dina[n]ti da voi et segnando le mire vostre cioè A • B • C • co[n] tre p[er]le ò tre paternoster.[2]

---

[1] Transcriber's note. Partially illegible because of the particular mark used for the first letter 'u'.

[2] Translator's note. Cfr Grayson: 'Questo medisimo potete fare con uno filo apiombinato, facendolo pendere dinanzi da voi e segnando le mire vostre [...] con tre perle [...] **come altre volte vi mostrai**'.

1r

If you wish,[3] by sight only, being at the head of a square, to measure how high that tower at the foot of the square is, do like this: Stick a spear in the ground, and fix it so that it stays straight up, and then move away from the spear by some distance, say six or eight feet, and then aim <your vision> at the top of the tower, directing your sight directly in line with vertical of the spear; there where your vision strikes the spear, put a bit of wax as a mark, and let this wax be called A. And then, without moving either your feet or your face from where you aimed at the top, aim lower down to the foot of that tower, and in like manner, where your sight strikes the spear, put another bit of wax, and let this second wax be called B. Finally, aim at some spot on the tower that you know and that you can easily measure with your spear when you are near the tower, such as perhaps the arch of the portal or some slit[4] or such like near the ground. And as you did when you aimed at the top and at the bottom of the tower, here too place a third bit of wax on your spear where your sight strikes it. This is a known value, and let this third wax be called C, as shown in the figure.

Ostilius   Riccius

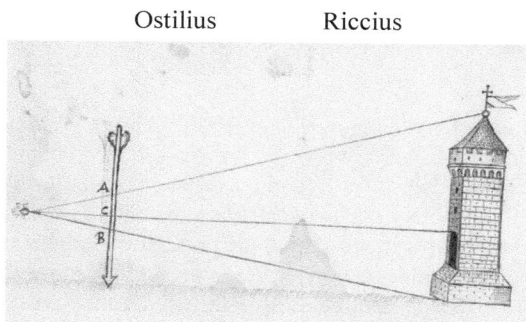

1v

I say that the number of times the part of the spear between wax B and wax C go into the other part of the spear that is between point A and point B, will be equal to the number of times that the lower part of the tower that you know will go into that part above that you do not yet know. And to make this doctrine clearer and more practical, here is an example in numbers for you. Let the tower be one hundred feet high, and the arch of the door in the tower 10 feet, you will find a similar ratio in the spear, that is, in the same way that the part of the tower, ten, goes into the greater and upper part nine times plus itself as one of the 10 parts of the whole tower, thus the part AC of the spear divided into nine parts will be such that it will receive nine times BC, and tenth of all AB. And so doing you will never err, as long as you never move your eye from its initial position. This same thing can be done with a plumb line, making it so it hangs in front of you and marking the points where your sight strikes the line, that is, A, B and C, with three pearls or three rosary beads.

---

[3] Translator's note. In this translation, additions integrated from the Grayson edition appear in bold type in square brackets [ ], while additions integrated by the translator for the sake of clarity appear in angle brackets < >.

[4] Translator's note. Original 'pertuso', modern 'pertugio', hole, slit, narrow opening.

Misurate in questo modo l'alteza d'una torre d[e]lla quale niuna parte a voi sara nota ma ben potete andar[e] insino al piè d[e]lla torre ficchate in terra come di sopra dissi uno dardo et scostatevi da questo dardo quanto vi par[e] et ponete l'occhio giu basso alla terra et indi mirate alla cima d[e]lla torre dirizando il vedere taglia il dardo • B • et questa cera postovi • C • et l'occhio vostro si chiami D • come nella faccia seguente vedete figurato:[5]

2 r

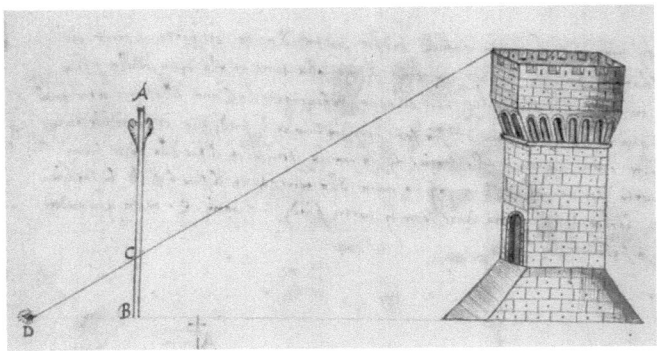

Dico che la parte del dardo quale sta fra • C et B • entra tante volte nella distantia quale sta infra • B • et [D][6] cioè infra l'occhio vostro et il pie d[e]l dardo quante volte l'altaza d[e]lla torre entra nella distanzia quale è fra l'occhio vostro et il pie d[e]lla torre è per exemplo: sia la torre alta piedi cento, et l'occhio vostro sia distante dal pie della torre piedi mille troverete nel vostro dardo ch[e] la mira risponde pure simile cioè come cento entra in mille dieci volte così • C • et B • entra in • d • et B • et secondo il numero saprete quante alteze della torre entrano i[n] tutta la distantia ch[e] sia fra l'occhio vostro et il piè della torre senza nissuno errore.[7]

et questo medesimo potrete pur[e] fare col filo segnato il punto • C • co[n] la sua perla Par ad alcuno più breve via tanto appressarsi alla torre ch[e] stando voi a diacer[e] in terra et tocchando co piedi il dardo fitto in terra come è detto di sopra la mira dritta alla cima d[e]lla torre batta nel dardo alto q[uanto] p[ro]p[rio] sia dal'occhio vostro a piedi et dicono il vero ch[e] tanto sarà dal pie d[e]lla torre a l'occhio vostro quanto dal medesimo pie fino alla cima alui[8] danno modi quali sono verissimi, et dicono:

---

[5] Translator's note. Cfr. Grayson: 'Ficcate in terra come di sopra dissi un dardo, e scostatevi da questo dardo quanto vi pare, e ponete l'occhio giù basso alla terra, e indi mirate la cima della torre, dirizzando il vedere **vostro per mezzo la dirittura del dardo, e lì dove il vedere** taglia il dardo **ponete una cera, e chiamasi la cima del dardo A e il piè** B, questa cera postavi C, e l'occhio vostro D, come qui vedete figurato'.

[6] Transcriber's note. Unreadable.

[7] Translator's note. Cfr. Grayson: '...troverrete nel vostro dardo che la mira risponde pure simile, cioè come cento entra in mille dieci volte, così C e B entra in DB **pur dieci volte. Adunque voi misurate quante volte CB entra in DB, e** secondo il numero saprete quante altezze della torre entrano in tutta la distanza che sia fra l'occhio vostro e il piè della torre senza niuno errore'.

[8] Transcriber's note. Probable transcription error for 'alcuni'.

Measure in this way the height of a tower of which no part is known to you, but of which you can easily arrive at the foot. Stick in the ground, as I said above, a spear, and move some distance away from the spear, and place the eye low down on the ground, and then aim **[at the top of the tower, directing your sight through the straight line of the spear and there, where the]** vision cuts the spear **[place a bit of wax. Let the top of the spear be called A, and the foot]** B, the wax you placed C, and let your eye be called D, as shown here.

2r

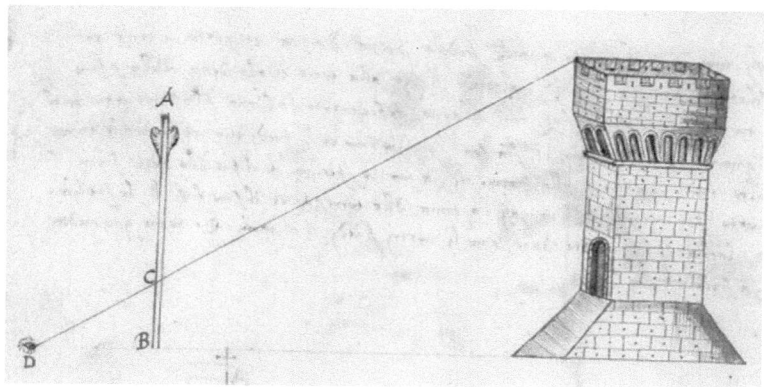

I say that the part of the spear that lies between C and B goes as many times into the distance between B and D, that is, between your eye and the foot of the spear, as many times as the height of the tower goes into the distance that is between your eye and the foot of the tower. And for example, let the tower be one hundred feet tall, and your eye be one thousand feet away from the base of the tower. You will find in your spear that the sight corresponds likewise, that is, as a hundred goes into a thousand ten times, so <the distance between> C and B goes into <the distance between> D and B ten times. According to the number you will know how many heights of the tower go into the whole distance that is between your eye and the foot of the tower without any error.

And you can do this same thing with a string, point C marked with a pearl. Some think that a shorter way is to go close to the tower, so that, when you lie down and touch the spear stuck in the ground with your feet, as I said above, the line of sight straight up to the top of the tower will strike the spear precisely as high as your eye is from your feet. And they are right, that as much as the base of the tower is from your eye, so that much it will be from the same base up to the top. Others give ways that are most true and useful, and say:

2v

Togli uno [specchio][9] ò più presto qualch[e] scodella piena d'aqua et posta in terra et
dischostati da essa sempre volgendo il viso alla torre et alla detta scodella p[er] fino ch[e]
tu vegga in quella sup[er]fine[10] dell'aqua rapresentata la cima d[e]lla torre et troverete
ch[e] quante volte lo spatio ch[e] sia fra l'occhio nuo[11] et li piedi tua et lo spechio tante
volte entra la torre nello spatio ch[e] sia fra lo spechio et il pie d[e]lla torre[12] sieni questo
per exemplo chiamasi la cima della torre • A • et il suo pie • B • lo spechio • C • lochio di[13]
• vostro dare[14] sono li vostri piedi si chiami • E • come qui vedete la pictura:

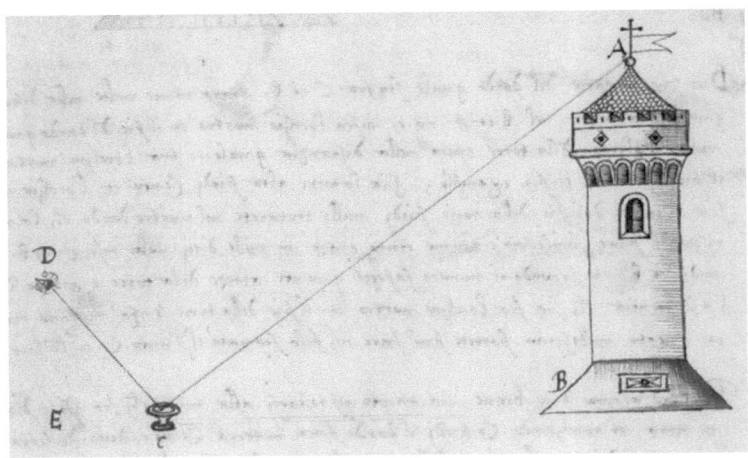

Dico che se • A • B sono pied[i] cento et • B • C sara piedi mille troverrete pari ragioni • C
et E et • d • E cioè ch[e] come cento entra in mille pur[e] X volte cosi • d • E entra in • C •
E volte pur dieci.

3r

Se volete misurar[e] le alteza duna tor[re] dove no[n] vi potete accostar[e] ma ben vedete il
suo pie et la sua cima vi conviene trovar[e] modo di sapere qua[n]to sia lo spatio ch[e] è
fra voi et il pie d[e]lla torr[e] p[er]ch[è] se sapete bene corre questo spatio anchora cio le
misure[15] sopra recitate saprete bene intender[e] sua alteza p[er] saper[e] questa distanzia ci
sarebbe modo qual porremo qui di sotto apunto a misurare ogni distanzia maxime quando
no[n] sia troppolontana p[er] misurar[e] le molto lontane vi darremo modo singulare.

---

[9] Transcriber's note. Missing word; Translator's note. Cfr. Grayson: 'Togli uno **spechio**,...'.

[10] Transcriber's note. Abbreviation and erroneous transcription for 'superficie'.

[11] Transcriber's note. Erroneous transcription for 'tuo'.

[12] Translator's note. Cfr. Grayson: '...troverrete che quante volte lo spazio che sia fra l'occhio tuo e'
piedi tuoi, **entra nello spazio che sia fra' piedi tuoi** e lo spechio, tante volte entra la torre nello
spazio che sia fra lo spechio e il piè della torre'.

[13] Translator's note. Cfr. Grayson: 'Chiamisi la cima della torre A e il suo piè B, lo spechio C,
l'occhio D, **e il sito** vostro dove sono e' vostri piedi si chimi E, come qui vedete la pittura'.

[14] Transcriber's note. Erroneous transcription for 'dove'.

[15] Translators's note. Cfr. Grayson: '...questo spazio, **allora con** le misure...'.

2v

Take a mirror, or more handily, a pan full of water, and place it on the ground, and move away from it, keeping your face turned towards the tower and said pan, until you can see the top of the tower reflected in the surface of the water. You will find that as many times the space that is between your eye and your feet **[goes into the distance between your feet]** and the mirror, then that many times the tower goes into the space there is between the mirror and the foot of the tower. Here is an example for you. Let the top of the tower be called A and its base B, the mirror C, the eye D, **[and]** your **[place]** where your feet are is called E, as you see in the figure.

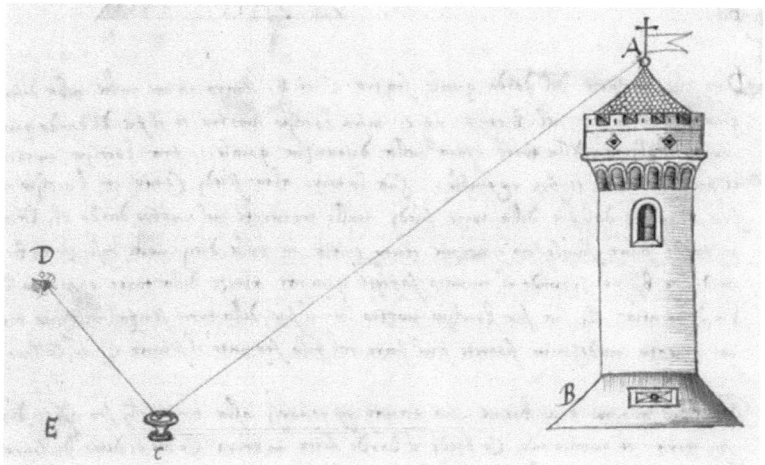

I say that if AB is a hundred feet, and BC is a thousand feet, you will find equal ratios between CE and DE, that is, as a hundred goes into a thousand 10 times, so DE goes into CE ten times as well.

3r

If you want to measure the height of a tower that you can't get close to, but you can see its base and its top well, it is best to find a way to know how much the space between you and the foot of the tower is, because if you know this space for sure, then from the measures given above you will be well able to understand how high it is. There is a way to know this distance that I will set forth here below, suitable for measuring any distance, especially when it isn't very far away. To measure those quite far way, we will give you a singular way.[16]

---

[16] Translator's note. This 'singular way' is given further on, in problem 17 (see fols. 14r-14v on pp. 56–62 in this present volume).

Misurate la largheza d'uno fiume sendo in su la ripa sua in questo modo ponetivi col pie in luogo piano et li fichate in terra uno dardo, come diremo di soppra, et chiamasi questo dardo A • B in questo dardo p[ro]p[rio] ponete allaltezza d[e]l'ochio vostro una cera et chiamasi questa cera • C • dipoi scostatevi da questo dardo • A B • quanto aprite le braccia et fichatevi uno altro dardo come di sopra et chiamasi questo secondo dardo • D • E et in questo • D • E ponete simile una cera proprio al'altezza d[e]l ochio vostro, et chiamasi questa cera • F • tenendo l'occhio giunto a questa cera • F • et mirate per dirittura d[e]l dardo • A • B • qual cosa nota di la dal fiume quale sia in sula ripa come sarebbe uno cespuglio ò qualch[e] luogo ò sasso et chiamasi questa cosa • G • et dove mirando il vostro veder[e] taglia il dardo • A • B • vi ponete una altra cera et chiamasi • H come qui a pie vedete la pittura

3v

Dicho ch[e] se misurate lo spatio fra la prima et la seconda cera d[e]l primo dardo cioè lo spatio • C • H • quante volte eglentri[17] fra lo spatio ch[e] sta fra l'uno dardo et l'altro cioè[18] CF tanto troverrete ch[e] H • B • entra in B • G • cioè lo spatio quale è nel primo dardo et il cespuglio qual voi mirasti eccovi lo exemplo[19] a numeri cioè:

Sia il fiume largo passi trenta et sia lo spatio • C • B • et simile lo spatio F • E uno passo il punto • H sara distante dal punto • C • tanto ch[e] lo pari intrerra in • F • C entra in C • F trenta volte F • E entra in E • G • pure trenta volte[20] ch[e] sara largo il fiume trenta volte quanto è da locchio vostro al pie. Eccovi una altra via più expeditiva se il paese dove voi siete sarrà piano fate come dicemo di sopra ponete dua dardi in terra et segnate tutto come dissi C • f • H • et pigliate la misura quanto sia fra • C • et H • et ponete una cera a questa medesima misura sotto f • nel dardo • D • E qual cera si chiami I • et poi ponete l'occhio vostro ch[e] tochi el primo dardo cioè A • B • proprio nel punto • C • et mirate per la dirittura d[e]lla cera I posta nel secondo dardo F • E et dove il veder vostro batte in terra la oltre a lunghi d[e]l dardo F • E • vi fate porre uno segno cioè uno sasso ò ch[e] vi pare, et chiamasi questo segno K come qui di sotto p[er] signio vi si mostra cioè:

---

[17] Transcriber's note. Contraction for 'egli entri'.
[18] Translator's note. Cfr. Grayson: 'Dico che se misurerete lo spazio fra la prima e la seconda cera del primo dardo, cioè **in AB** lo spazio CH, quante volte egli entri fra lo spazio che sta fra l'uno dardo e l'altro...'.
[19] Transcriber's note. Erroneous transcription for 'exemplo'.
[20] Crf. Grayson: '...el punto H sarà distante dal punto C tanto ch'ello pari entrerà in FC **tanto volte quante entra HB in BG, cioè trenta volte, e più se HC** entra in CF trenta volte, FE entra in EG pure trenta volte,...'.

Measure the width of a river, when you are on its bank, in this way. Place yourself with your feet in a flat place, and stick a spear into the ground there, as we said above, and let this spear be called AB. On this spear, precisely at the height of your eye, make a mark with wax, and let this wax be called C. Then move away from this spear AB by as much as you can open your arms wide, and there stick another spear like above, and call this second spear DE; and on this DE make a similar mark with wax at the height of your eye, and let this wax be called F. Keeping your eye fixed on this wax F, and aiming along the straight line of spear AB at something known[21] on the other side of the river, such as a bush or a place or a stone, and let this thing be called G; and where your line of sight cuts spear AB, place another bit of wax and let it be called H, as you see in the figure.

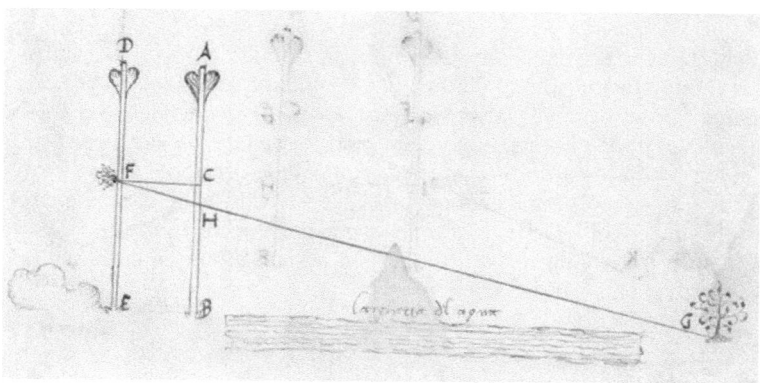

3v

I say that if you measure the space between the first and the second wax of the first spear, that is, how many times the space CH goes into the space that is between one spear and the other, that is, CF, you will find that that is the number of times that HB goes into BG, that is, the space that is between the first spear and the bush you aimed at. Here is an example in numbers for you.

Let the river be thirty paces wide, and the space CB and likewise the space FE one pace; point H will be as far from point C such that it goes into FC [equally as many times as HB goes into BG, that is, thirty times; further, if HC goes into CF thirty times, FE goes into EG] thirty times as well, that is, the river will be thirty times as wide as it is from your eye to the foot. Here is another very expedient way. If the place where you are is flat, then do as we said above. Place two spears in the ground and mark them as was said CFH, and measure how much there is between C and H, and place a wax at that same measurement below F on spear DE, which wax is called I. And then place your eye so that it touches the first spear, that is AB, precisely in point C, and aim along the straight line of wax I placed on the second spear FE, and where your sight strikes the [flat] ground there beyond spear FE, place a sign, a rock or something, and call this sign K, as you can see drawn below, that is,

---

[21] Translator's note. Original 'qual cosa nota', i.e., something known or identifiable.

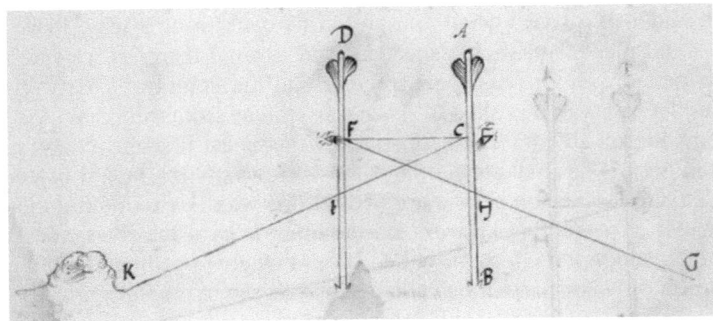

4r

Troverete ch[e] tanto sarà dal segno • K • p[er] in sino al dardo • A B • dal segno G quale sta di la dal fiume p[er] in sino al dardo • D • E • misura certissima ma questa che[22] segno sara piu maravigliosa, bench[é] la sia alquanto laborosa cio intendere Se vedete d'una torre solo la sua cima et nulla altra parte et volete saper[e] quanto sia alta fate così. Ponete come è ditto di sopra il vostro dardo fitto in terra et ponete le orechio[23] a terra et mirate la cima d[e]lla torre et segnate co[n] una cera dove il vedere vostro batte et chiamasi il dardo • A • B la cima d[e]lla torr[e] • C • il punto dove ponesti l'occhio in terra • D • la cera ch[e] ponesti nel dardo • E • fatto questo tiratevi piu adrieto et simile da basso mirate la detta cima d[e]lla torre et chiamassi G come p[er] pictura qui a pie vedete:[24]

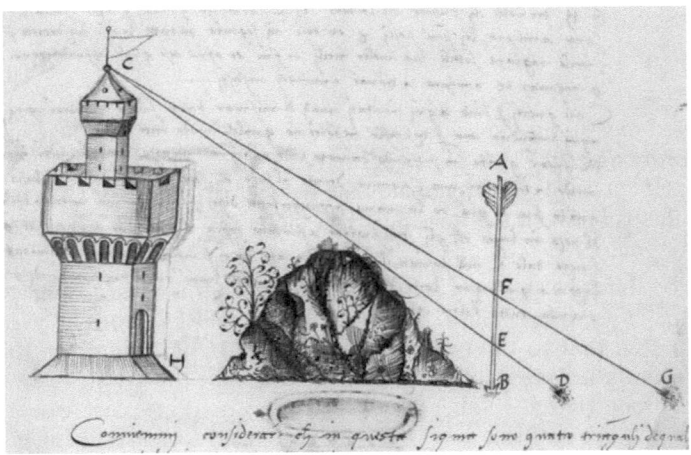

---

[22] Transcriber's note. Word difficult to interpret because of a sign used for the 'h' that is not found anywhere else in the text.
[23] Translator's note. Cfr. Grayson: '...e ponete l'**occhio** a terra...'. In any case, the transcription is meaninful; in fact, if you place your ear on the ground, you will be in the correct position to carry out the operation described
[24] Translator's note. Cfr. Grayson: 'Fatto questo, tiratevi più adrieto, e simile da basso mirate la detta torre, **e segnate dove testé batte el vostro vedere nel dardo, e chiamisi questa seconda cera F, e dove ponesti l'occhio** si chiami G, come qui vedete dipinto.

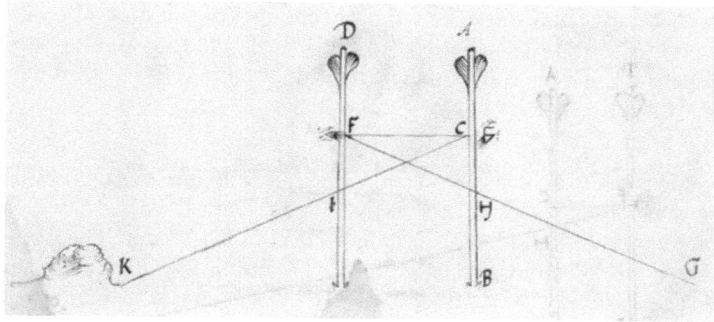

4r

You will find that as much as it is from sign K up to spear AB, is as much as from sign G that is on the other side of the river, up to spear DE; a most certain measure. But this which follows is even more marvellous, even though it is so much more laborious to understand. If you can see only the top of a tower and none of its other parts, and you want to know how high it is, do like this. Place, as said above, your spear firmly in the ground, and put your ear[25] to the ground and aim at the top of the tower, and mark with a wax where your <line of> sight strikes, and let the spear be called AB, the top of the tower C, the point where your eye[26] is placed D, the wax mark on the spear E. Having done this, pull back, and similarly from low down aim at said top of the tower, and **[mark where your sight just strikes the spear, and let this second wax be called F, and where the eye is placed]** be called G, as you can see drawn.

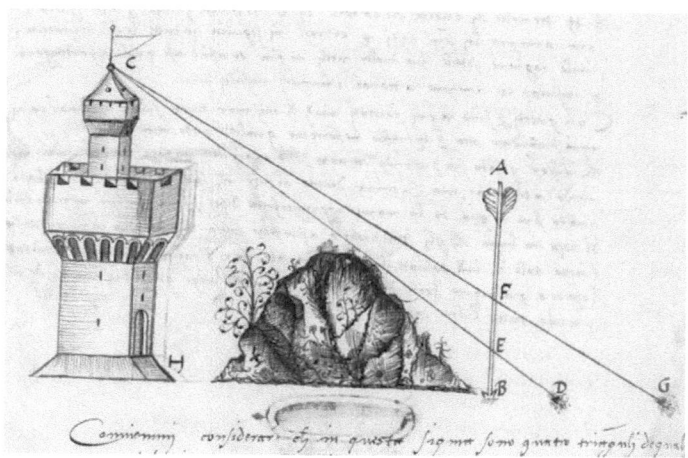

---

[25] Translator's note. Should be 'put your **eye** to the ground' (as shown in the figure).
[26] Translator's note. 'Eye' is correct here.

Convienvi considerar[e] ch[e] in questa figura sono quatro tria[n]guli de quali questi duoi sono a voi noti cioè F · B · G maggiore et laltro E · B · D minore p[er] questi verrete in cognitione di tutti e trianguli et maxime chiamati luno · C · H · G · laltro · C · H B · et voi intendrete per [recitati][27] di sopra ch[e] come la linea · D · B · .

4v

nel suo trangulo[28] così la linea G • H nel trangulo maxime risponde alla linea H • C adunque misurate per questa ragione et comparatione qua[n]te volte • D • B • entri in E • B • quale pogniamo per exemplo facile ch[e] lentri dua volte segnata[29] che C • H sia dua tanti quanto H • D • et più misurate quante volte B • G entra in B • F • quale metto caso ch[e] entri ne[30] ne seguita ch[e] C H sia il terzo di • H • G • et simile seguita ch[e] da D H sono due et da G • H son tre numeri no[n] sapete questo numero quanta quantità sia, seglie[31] braccia o passi ò ch[e] etrovi el modo se D • H son due et H • G • son tre seguita ch[e] H • G • avanza H • D • d'uno et quelo ch[e] glie avanza è H • G[32] • adunque epso • D • G • è uno terzo misurate questo D • G • qual sara passi dieci tutto H • G • sara trenta diqui argume[n]tate in questo modo se la torr[e] C • H entra in tutto qu[e]sto spatio H • G • tre volte et D • G • è il terzo et simile entra lui in tutto G • H tre volte chi dubita ch[e] la torr[e] • H C • è lunga qua[n]to è questo spatio sarà ancora lei pur passi x[33] et così vi seguita in tutte le cose misurerete simile ragione sottile ma molto utili e più e a più cose quale appartengono a misurar et anchora a trovar[e] e numeri nascosti.[34]

Con questi p[er] sino a qui recitati modi di misurar[e] potete simile misurar[e] ogni profondità ma p[e]r exemplo ne porremo qualch[e] modo certo.

Misurar[e] quanto sia profondo cavato fino a l'acqua un pozo solo col veder[e] in questo modo a traversar una can[n]uccia dentro al pozo ch[e] stia ferma dase giu basso qua[n]to piu in giu co[n] la mano aggiunger[35] potete dipoi ponete l'occhio vostro a l'orlo d[e]l pozo in luogo ch[e] gli stia proprio a piombo sopra il capo d[e]lla can[n]uccia e sia luogo tale ch[e] indi possiate il fondo d[e]l cavato cioè p[e]r infino a l'acqua e mirate laggiu a quel'acqua l'orlo d[e]lla superpicie in quel luogo quale proprio risponde a piombo tutto l'altro capo

---

[27] Transcriber's note. Word difficult to read; Translator's note. Cfr. Grayson: '...e voi intenderete **pe' modi recitati** di sopra...'.

[28] Translator's note. Cfr. Grayson: '...la linea DB **risponde alla linea EB** nel suo triangulo...'.

[29] Translator's note. Cfr. Grayson: '...due volte, **seguita** che...'..

[30] Transcriber's note. Apparent transcription error with the repetition of 'ne' and the omission of the word 'tre'. Cfr. Grayson: '...entri **tre**, seguita che...'.

[31] Transcriber's note. Contraction for 'se gli è'; Translator's note. Cfr. Grayson: '...**s'egl'è** braccia o passi...'.

[32] Translator's note. Cfr. Grayson: 'DG', instead 'HG'.

[33] Translator's note. Cfr. Grayson: '...chi dubita che la torre HC è lunga quanto è questo spazio **DG? Ma questo spazio DG è dieci; adunque la torre uguale a questo spazio** sarà ancora lei pur passi dieci'.

[34] Translator's note. Cfr. Grayson: '...numeri **ascosi**'.

[35] Translator's note. The word order in this sentence differs significantly from Grayson but the meaning is the same. Cfr Grayson: 'Traversate una cannuccia dentro al pozzo giù basso quanto più potete giugnere con la mano, e fermatela che la vi stia ben ferma da sè'.

You do well to consider that there are in this figure four triangles, of which these two are known to you, that is, the larger one FBG and the smaller one EBD. By these you will come to know all the largest triangles, one called CHG and the other CHD, and you will understand by the ways described above that as the line DB

4v

[corresponds to the line EB][36] in its triangle, so GH in the largest triangle corresponds to line HC. Thus, measure by this ratio and comparison how many times EB goes into DB.[37] Let us say, for simplicity's sake, that it goes in two times, it **follows** that HD will be two times as much as CH.[38] And further, measure how many times BF goes into BG,[39] say it's the case that it goes in three <times> it follows that CH is the third <part> of HG. And similarly it follows that from DH there are two and from GH there are three numbers. You don't know how many quantities this number is, be they *braccia*[40] or paces or whatever. Here is how you do it. If DH are two and HG are three, it follows that HG is greater than HD by one, and that by which it is greater is **DG**. Thus this DG is a third. Measure this DG, and if it is ten paces, then all of HG will be thirty. About this you should reason in this way. If the tower CH goes into all of this space HG three times, and DG is a third and it likewise goes into all of GH three times, who can doubt that tower HC is as long as this space; it will also be ten paces. And thus you will go about measuring all things, ratios that are similarly subtle but very useful for more and more things that appertain to measuring and also to finding hidden numbers.[41]

With the ways of measuring described up to now, you can similarly measure all depths, but for example, we will set forth some sure ways.

Measure, only by sight, how deep a well goes down to the water in this way. Lower a thin reed[42] into the well as far down as you can with your hand, and fix it there so that it stays by itself. Then put your eye at the rim of the well so that it is directly above the end of the reed, and let this place be such that you can see the bottom of the well, that is, down to the level of the water, and aim <your sight> down there towards the water at the edge of the surface in that place that corresponds directly under the other end

---

[36] Translator's note. With or without this addition from Grayson, this sentence is problematic, since quantities from triangle *EBD* are claimed to be in the same ratio with quantities from the non-similar triangle *CHG*. Perhaps the inherent confusion is why the manuscripts differ here.

[37] Translator's note. It is clear from a mathematical point of view that *DB* and *EB* are interchanged here, as they are in Grayson, so they are switched in the translation.

[38] Translator's note. It is clear from a mathematical point of view that *CH* and *HD* are interchanged here, as they are in Grayson, so they are switched in the translation.

[39] Translator's note. It is clear from a mathematical point of view that *BG* and *BF* are interchanged here, as they are in Grayson, so they are switched in the translation.

[40] Translator's note. A 'braccio' (plural 'braccia') is a measure of length, roughly about 60 cm.

[41] Translator's note. 'Hidden numbers', i.e., measurements of lengths or quantities not directly accessible.

[42] Translator's note. 'Cannuccia', straw (like a drinking straw) or reed.

5r

d[e]lla vostra can[n]uccia et chiamasi questa can[n]uccia il capo lontano da voi • A l'altro capo presso a voi • B • l'occhio vostro • C el bassa d[e]l pozzo sopra d[e]l'acqua • D

fatto questo mirate il luogo d[e]l'agua detto • D • et dove il veder[e] vostro batte nella can[n]uccia ponete una cera p[er] segno et chiamasi • E • come qui vedrete fignato:

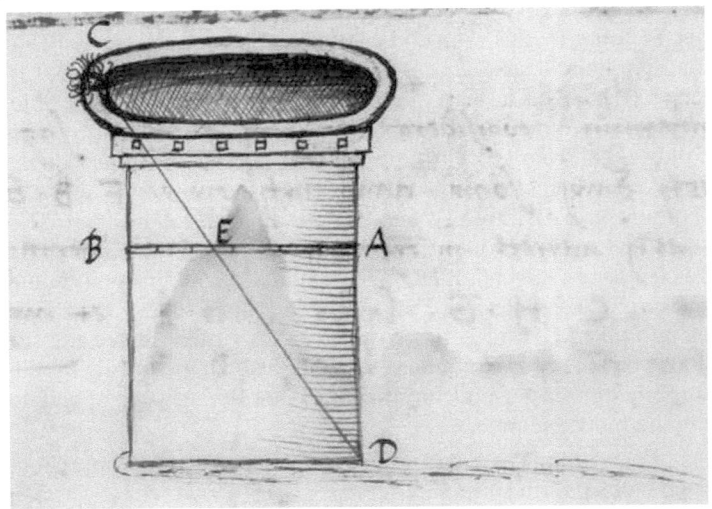

Dicho ch[e] quante volte • E • B • entra in B • C • cioè quante volte lo spatio ch[e] sta nella can[n]uccia infra • E • B · entra nella parte d[e]l pozo quale sta fra le occhio vostro fino al capo della canuccia posto a piombo sotto l'occhio vostro tante volte A • B • cioè tutta la canna misura tutto il profondo d[e]l pozo ecchovi lo exemplo:

Sia profondo il pozo b[raccia] 21[43] sia • A B • cioè tutta la canna et la largheza d[e]l pozo b[raccia] 3 enterra aduncq[ue] • A • b • volte septe in tutta la profondita così troverette misurando come dissi ch[e] E • B • entra in B • C tante dette ch[e] sono la can[n]uccia qual misurato il vostro pozo no[n] mi extendo qui in misurar[e] queste profondita però ch[e] voi a vostro ingegno per vostra similitudine tutto comprenderete.

Ma no[n] preterni qui certo modo posto dagli scriptori antichi apto a misurare una profondità d'una acqua molto cupa qual sarebbe la valle di odrie o simile anchora più profondo:

5v

Se volete misurar[e] la valle quanto sia profonda qual no[n] si trovi fondo co[n] lo schagliando ne co[n] molte fune fate cosi habbiate uno vaso apto a tener[e] acqua sia bossolo ò taza ò ch[e] vi pare fategli nel fondo un piccolo p[er]tuso et habiate una galla di quercia et apichatevi uno ferretto minuto fatto simile a una figura di abbaco quale importi •

---

[43] Transcriber's note. Difficult-to-read Arabic number; Translator's note. Cfr. Grayson: '...braccia **ventuno**;...'.

5r

of your reed. Let the end of this reed that is far away from you be called A, and the other end near you B, your eye C, and the low end of the well at the water level D.

This done, aim <your sight> at the place at the water level called D, and where your sight strikes the reed, make a mark with wax, and let this be called E, as you see drawn here:

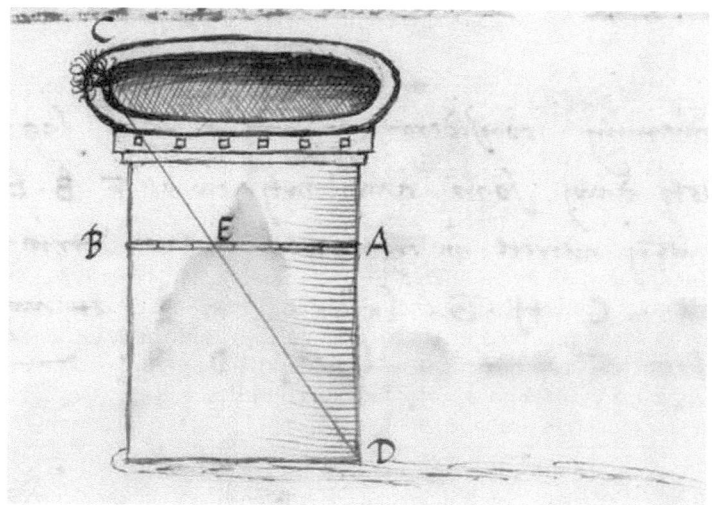

I say that as many times as EB goes into BC, that is, as many times the space that there is in the reed between EB goes into the part of the well that is between your eye up to the top of the reed placed directly under your eye, that many times AB, that is, the whole reed, measures the whole depth of the well. Here is the example for you.[44]

Let the well be 21 *braccia* deep; let AB, that is, the reed, and the width of the well 3 *braccia*. Thus AB will go seven times into the whole depth. Thus you will find, measuring as I have described, that EB goes into BC as many times[45] as there are reeds that measure your well. I won't go into measuring these depths here, because I know that you have the wits to understand all things that are similar to this. But here I won't omit a certain way described by the ancient writers, intended to measure the depth of very dark water, like the basin of the Adria or others that are even deeper.

5v

If you want to measure how deep the basin is whose bottom can't be found with either a sounding line or many chains, do it this way. Take a vessel suitable for holding water, be it square or round or whatever; make a small hole in the bottom. And take an oak-gall, and

---

[44] Translator's note. In Gal. 10 the drawing shown above is actually misplaced. Following this paragraph there is a large blank space as for a figure, but the figure appears at the bottom of the page 4v.

[45] Translator's note. Both Gal. 10 and Grayson have '...tante dette' instead of the more often used 'tante volte', many times.

5[46] · et questo ferretto quel gambo maggior[e] ficchatelo in quella galla p[er] fino alla meta sua l'altro mezo ava[n]ti fuori d[e]lla galla habbiate piombini apti di presso [quanto][47] vi pare che sforzeno la galla vostra a ire al fondo d[e]lla acqua or questi piombini sieno facti in questa forma quale a pie vedete dipinta simile il vaso et simile la galla:

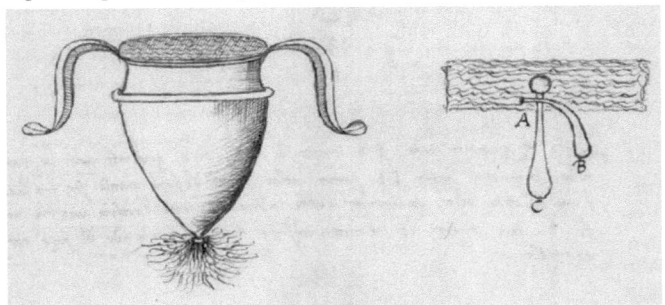

Appichate uno di questi piombini alla vostra galla come vedete la pittura et andate in luogo dove a voi sia noto et misurate li quanto sia il fardo[48] de l'acqua et qui impiete il vostro vaso di acqua et sia lacqua pura et pesate l'acqua co[n] tutto il vaso bene apunte quante l[i]b[re] ò de grani sia[49] fatti questi apparechiame[n]ti a uno tratto lasciate ire[50] la galla col suo piombo in acqua et inpiamo a pur[e] l'acqua ch[e] l'esca d[a]l vaso[51] qui la galla tirata dal piombo andar p[er] fino al fondo giunto ch[e] sarà il piombo il capo

6r

suo chiamato • C[52] • tochera prima il terreno et fermerassi et il capo • B • simile declinera a terra et in di la coda applicata a langula d[e]l ferruzo si distorra dal luogo suo et la galla libera rivoltera suso ad altro.

Siate presto et dividete col dito ch[e] nulla più aqua esca d[a]l vaso et pesate quanta acqua vi resta et quanta vi ne mancha et notate in questo tempo ch[e] la galla ando giu et ritorno tante br[accia] quanta acqua si verso no[n] mi extendo più credo ch[e] assai comprenderete ch[e] co[n] questa misura vi sara facile al misurar[e] in profondo d[e]l oceano pure ch[e] l'acqua no[n] sia corrente.

Con queste simili ragioni et vasi si fanno horologii assai gusti p[er] misurar[e] il tempo a ore et meze ore et simile a molte cose sono acomodate.

Insom[m]a ogni cosa dove sia alcuno moto[53] sara apta a misurar[e] il tempo et di qui sono tutti gli orologii fabricati come quegli dove certi pesi cercono posarsi in terra quali sono contrapesi la polver[e] l'acqua et simili.

---

[46] Transcriber's note. Difficult-to-read Arabic number; Translator's note. Cfr. Grayson: '...quale importi **5**,...'.

[47] Transcriber's note. Illegible; Translator's note. Cfr. Grayson: '...**atti di peso quanto** vi pare,...'.

[48] Transcriber's note. Apparent transcription error for 'fondo'; Translator's note. Cfr. Grayson: '**fondo** dell'acqua,...'.

[49] Translator's note. Cfr. Grayson: 'quante **once** e quanti grani'.

[50] Transcriber's note. The 'r' has been the object of a correction, making it difficult to read.

[51] Translator's note. Cfr. Grayson: '...e **insieme aprite** l'acqua ch'ell'esca del vaso'.

[52] Translator's note. Cfr. Grayson: 'el capo suo **segnato** C'.

stick to it a minute wire made similar to an abacus figure that stands for 5, and stick the larger leg of this little wire into the gall up to its halfway point; the other half will stick out of the gall. Take as many plumbs suited to act as weights as you wish, so that they force your gall to go to the bottom of the water, and let these plumbs be made in the form that you see drawn here, and the vessel and the gall similar to the drawing as well.

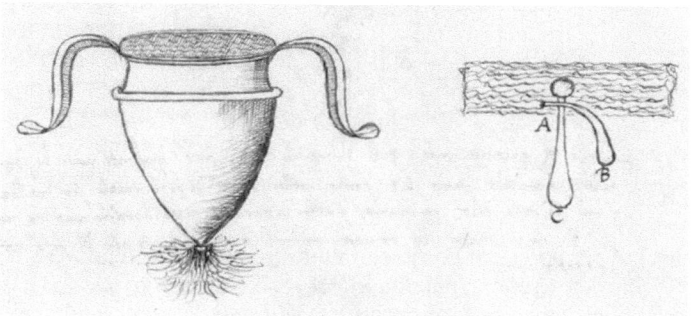

Attach one of these plumbs to your gall as you see in the drawing, and go to a place you know and measure there how deep the water is, and here fill your vessel with water; and let the water be pure,[54] and weigh the water along with the vessel carefully to know how many pounds or grains it weighs. This equipment being made, in a <single> stroke, let the gall with its plumb go in the water, and at the same time open the water so that it goes out of the vessel. Here the gall dragged down by the plumb will go down to the bottom. When the plumb has touched down, its end

6r

called C will first touch the terrain and stop, and end B will likewise go down to the ground, and then the tail A attached to the angle of the little wire will be diverted from its place, and the free gall will return up to the surface. Be quick and close <the vessel> with your finger so that no more water goes out of it, and weigh how much water remains and how much has gone out, and note that it took this much time for the gall to go down and come back up for as many *braccia* as water went out. I won't go on; I believe that you understand quite well that with this measure you can easily measure the depth of the ocean, as long as the water is not flowing.[55]

With similar reasonings and vessels are made clocks that are very accurate; and for measuring time into hours and half hours and so forth, and many other things.

In short anything where there is any motion can be used to measure time, and on this <principle> are all clocks built, such as those where certain weights strain towards the ground, with soil, water and the like used as counterweights.

---

[53] Translator's note. Cfr. Grayson: 'Insomma ogni cosa **in cui** sia alcun moto,...'.
[54] Translator's note. 'Pure' water, i.e., water that doees not contain any silt or soil.
[55] Translator's note. 'Not flowing', i.e., as long as there is no current to carry the oak gall away.

Anchora si fa horologii col fuoco et co[n] laria hanno certi stoppini di talco et notano quanto peso d'olio ardono p[er] hora et cosi al tempo accendono el loro stoppino et assai risponde giusta questa ragione, lo horologio quale si fa a vento è cosa molto gioconda p[er]ch[é] questa è una fonte facta ch[e] posta in tavola certo spatio di tempo ella butta acqua in aria p[er] forza d'aria quale sputa fuori et sta in questo modo cioè:

Voi havete uno vaso lungo tre palmi ò quanto piace a voi d[e]l quale e labri di sopra si chiamano A • B• et il fondo di sotto si chiami • C • D a questo vaso voi ponete dua altri fondi altro l'uno da l'altro una stranna[56] et chiamasi el primo sopra posto fondo • E• F• et il secondo cioè quello ch[e] è più presso al labro di sopra si chiami H • G • questi fondi et questo vaso sieno bene stagnati ch[e] nulla p[er] alcuno luogo respiri nel fondo • G H cioè nel supremo fatevi uno fondo et stagnatevi una canna Bersa quale vi stia drento a pendicolo[57] et passi sotto questo fondo G • H • p[er] in fino al fondo • E • F • appresso et di[58] sopra ava[n]zi fuora de labri • A • B • et chiamasi questa cannella • I • K •

6v

Simile fate ch[e] vi sia uno altro p[er]tuso in questo medesimo fondo G • H et simile sotto questo a pendicolo sia anchora nel fondo • E • F • uno foro et p[er] questi duoi fori ponetevi un'altra can[n]ella p[er]forata ch[e] passi l'uno da laltro fondo cioè • G • H et E • F • et vada il capo di questa cannella in sino giù p[re]sso il fondo • C • D • et da l'altro suo sopra rimanga uvale al fondo • G • H[59] et chiamasi questa canella il capo di sopra • L •• et il fondo • M[60] • Item nel fondo E • F • sia un[o] foro drentovi fitta una cannella quale il capo suo a basso sia uvale a decto fondo • E • F[61] • et chiamasi O • et il capo alto sia p[er] infino presso al fondo • G • H et chiamasi N • saranno adunque[62] come qui vedete la pictura tre fondi luno sopra l'altro cioè • C • D • et • E • F et • G •• H • et tre canelle la prima I • K • qual sola possa il fondo • G H • et p[er] l'altro di tutto cioè EF et l'ultima cannella • N • O • qual solo passa anchora[63] il fondo EF[64] aggiungnete al fondo G • H • uno foro sanza niuna can[n]ella p[er] il quale si possa il vaso empier[e] d'acqua come qui diremo[65] et chiamasi el decto foro · [P][66] come qui apie vedete formato

---

[56] Translator's note. Translator's note. Cfr. Grayson: 'A questo vaso voi ponete due altri fondi **alto** l'uno dall'altro una **spanna**...'.

[57] Translator's note. Gal. 10 has 'fondo', level, instead of 'foro', hole; cfr. Grayson: 'Nel fondo GH, cioè nel supremo, fatevi un foro, e stagnaevi una canna busa quale vi stia entro a perpendicolo,...'.

[58] Transcriber's note. Word illegible.

[59] Translator's note. Cfr. Grayson: '...e dal lato suo di sopra rimango **uguale** al fondo GH;...'.

[60] Translator's note. Cfr. Grayson: '...e chiamisi questa cannella, el capo di sopra L., di sotto M.

[61] Translator's note. Cfr. Grayson: '...quale el capo suo abasso sia **uguale** al detto fondo EF,...'.

[62] Transcriber's note. Difficult to read.

[63] Transcriber's note. The beginning 'a' is illegible.

[64] Translator's note. Here Gal. 10 contains a significant passage that is not in Grayson; cfr. Grayson: 'Saranno adunque, come qui vedete la pittura, tre fondi l'un sopra l'altro, cioè CD e EF e GH, e tre cannelle: [...] IK quale passa un fondo EF e aggiunge a fondo GH, [...] NO quale [...] passa el fondo EF.

[65] Translator's note. Cfr. Grayson: '...come **più giù** diremo'.

[66] Transcriber's note. Illegible. Cfr. Grayson: '...detto foro **P**, ...'.

Further, clocks can be made with fire and with air. They have certain wicks of talc, and keep track of how much the weight in oil is burned by the hour, and thus when required they light their wick, and quite accurately they correspond to this reasoning. The clock that is made with the wind is a very jolly thing, because this is a fountain made so that, placed on a table for a certain amount of time, it spouts water into the air by force of the air, which it spurts out and is made thus:

You have a vessel three palms wide or as wide as you want, of which the upper rim is called AB, and the lower bottom is called CD. In this vessel place two more levels a span apart, and let the first superimposed level from the bottom be called EF; the second, that is, the one above that is closer to the upper rim, is called HG. These levels and this vessel should all be made quite watertight[67] so that no air at all can enter. At level GH, that is, the upper one, make a hole, and fix there and make watertight a little hollow tube that stays perpendicular, and pass below this level GH down close to the level EF, and above it should stick up beyond rim AB; let this little tube be called IK.

6v

Similarly make another hole in this same level GH, and similarly straight under this let there be in level EF a hole; and through these two holes place another hollow tube that goes from one to the other level, that is GH and EF, and the end of this tube should go down close to the level CD, and its upper part should remain equal to the level GH; and let the upper end of this tube be called L, and the bottom M. Finally, in level EF let there be a hole with a tube fit into it, with its lower end equal to said level EF, and let it be called O; the upper end goes up close to level GH, and is called N. Thus there will be, as you see in the drawing, three levels one above the other, that is, CD and EF and GH, and three tubes: the first IK that only passes through level GH; LM that goes through level EF and reaches level GH; and for the other entirely, that is, EF, and the last tube NO which only goes again through level EF. Add to level GH a hole without any tube that can be used to fill the vessel with water, as we will describe here, and let this hole be called P, as you see drawn here below.

---

[67] Translator's note. The verb 'stagnare' used here can mean either 'solder or tin-plate', or 'render airtight or watertight'.

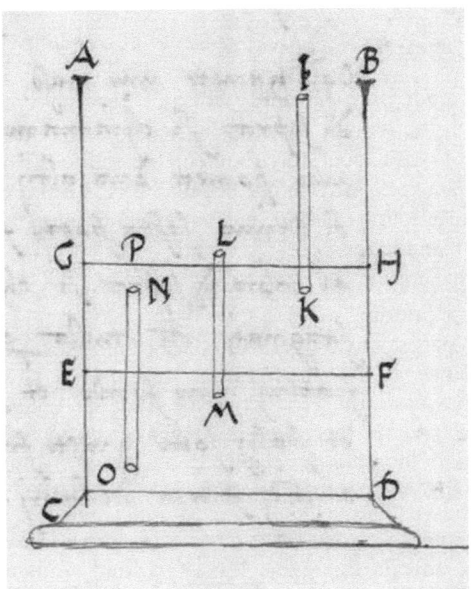

Eimpiete d'acqua p[er] il pertuso • P quella parte ch[e] sta fra il fondo G • H et il fondo • E F et turate bene il detto buso • P • ch[e] nulla piu acqua entri ne esca dipoi turate la boccha • L • della cannella M • N[68] et empiete il vaso d'acqua sopra la parte che sta fra e labri • A • B • et il fondo • G • H qua[n]to tutto sarà empiuto sturate la boccha • L della cannella L • M lacqua gira nelle parti giu fra • E F et C • D • qualle empiendosi l'acqua spugnera l'aria ch[e] verra et mand[e]ralla p[er] la cannella N • O • nella parte d[e]l vaso fra • E • F et G • H • indi l'aria pignera l'acqua p[er] la cannella I • K et quanto vi sara aria tanto durera

7r

il suo impeto a mandar[e] fuori lacqua giuocho molto dilectende et bello

Nel numero d[e]lli horologii sono optimi et certissimi quelli ch[e] notano il moto d[e]l sole et d[e]lle stelle et questi sono molti et varii come ostrolobio el quadra[n]te le armille et quelli anelli portati quali io soglio fare et simili et [di][69] questi la loro ragione è da molti scripta et è cosa prolissa. Ma qua[n]to e sia apto a questi ludi quali io raco[n]to sara questo quasi tutti si releggono co[n] la linea d[e]l mezodi[70] pero ch[e] ella è piu facta et più coequabile ch[e] termine alcuno ch[e] sia nel cielo adunq[ue] dico se volete trovare in ogni paese qual sia proprio il mezo fate così:

---

[68] Translator's note. Cfr. Grayson: '...cannella **LM**'.
[69] Transcriber's note. Illegible. Cfr. Grayson: '...E **di** questi ...'.
[70] Translator's note. Cfr. Grayson: '...quasi tutti si **regolano** con la linea del mezzodì,...'.

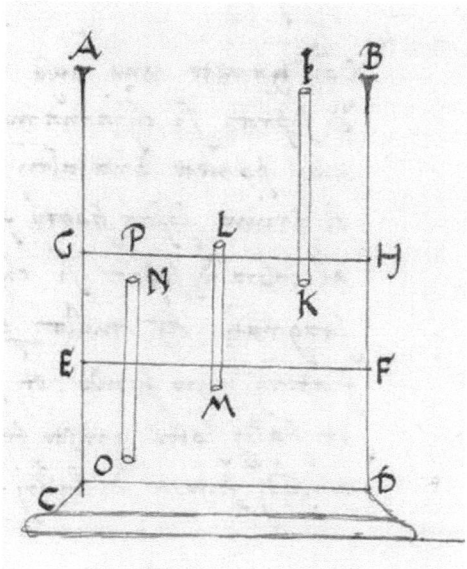

Fill with water through hole P that part that is between level GH and level EF, and carefully plug said hole P so that water can neither enter nor exit. Then plug the mouth L of tube LM, and fill the upper part container of water between rim AB and level GH. When all is ready, unplug the mouth L of tube LM. The water will swirl into the lower part between EF and CD which, filling with water, will push out the air that there was in it and send it out through tube NO into the part of the vessel between EF and GH. Then the air will push the water through tube IK, and as much air as there is, that much will its impetus continue

7r

to push the water out: a most delightful toy.

Among the number of clocks there are excellent and most accurate ones that note the motion of the sun and the stars, and these are many and various such as the astrolabe, the quadrant, the armillaries, and those portable rings that I make, and the like. And explanations of these have been written by many, and at length. But what is apt for these games[71] that I am telling about, is this: almost all are read from the noon line, because this is the most correct and most coequable of all terms that there are in heaven. Thus I say that if you want to find in any country where its own noon <line> is, do like this.

---

[71] Translator's note. This is the first of two instances of the word 'ludi' (plural for 'ludo') from which Alberti's title arises. 'Ludo' (plural, 'ludi') is literally translated as 'game', but also means 'amusement' or 'diversion'.

Ficchate in terra il vostro dardo in luogo piano[72] come di sopra ch[e] gli stia ben ritto et quando è sia dopo desinare inna[n]ti nona habiate uno filo legatelo intorno al dardo in terra[73] terminate il filo et fate girando uno arculo intorno al dardo in terra sarà dunq[ue] il ferro di questo dardo entro a questo circulo et chiamasi • A • dove proprio finisce l'ombra del dardo in sul circulo si chiami B · lascate stare così il dardo et in sul puncto B • • lascate stare così il dardo et in sul puncto B • ficchate uno stecho poi indi a una hora tornate et vedrete l'umbra del dardo battere alirave[74] aspectate chella proprio aggiunga a tocchar[e] il vostro circulo et segnate co[n] unaltro steccho questo luogo quale sara piu verso dove si leva il sole et chiamasi questo secondo steccho • C[75] • Come apie vedete la similitudine per disegno:

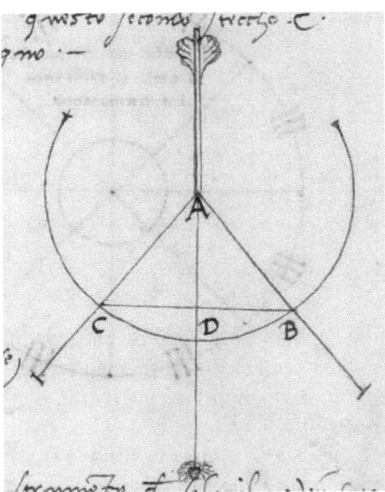

Dividete la linea • B • C cioè la distanza quale sta fra l'uno steccho et l'altro in dua parte eguale et chiamasi D • et dal p[un]to A • tirate uno filo drento a questo circulo a questo • D[76] • questa diritina proprio mira il mezodi[77] co[n] questo potrete porre quadra[n]ti dal sole iusti et ogni simil cose.

Per conoscere l'hore d[e]lla notte sanza altro strume[n]to ch[e] solo il veder[e] fate così. Notata la sera quando appariscono le stelle dove sia tramo[n]tana stella assai nota et ponete me[n]te sopra a quello al torre ò camino co[n] simile ella risponda sendo voi in questo certo luogo et notate di tutte le stelle sono circa

---

[72] Translator's note. Cfr. Grayson: 'Ficcate in terra **in luogo piano il vostro dardo,...**'.
[73] Translator's note. Cfr. Grayson: '...legalo **a piè di questo dardo. e proprio dove finisce l'ombra al sol di questo dardo,...**
[74] Transcriber's note. Probable erroreous transcription of 'altrove'. Translator's note. Cfr. Grayson: '...battere **altrove**'.
[75] Translator's note. Cfr. Grayson: '...e chiamisi questo [...] dardo **segno** C, come **qui** vedete la similitudine [...]'.
[76] Translator's note. Cfr. Grayson: '...e dal punto A **entro del circulo** tirate un filo a questo D'.
[77] Translator's note. Cfr. Grayson: 'Questa dirittura proprio mira il mezzo dì **in quel luogo**'.

Stick your spear in a flat place in the ground, like above, so that it is quite straight; and <at an hour> between the morning meal and noon,[78] take a cord, tie it around the spear in the ground, **[and right where the shadow of the sun ends on the spear],** cut the cord and turning it make a circle about the spear in the ground. Thus the embedded arrow of this spear will be the centre of this circle, and let it be called A. Right where the shadow of the spear ends up on the circle is called B. Let the spear stay that way and in point B put a stick. Then come back in an hour; you will see that the shadow of the spear strikes in another place. Wait until it precisely comes to touch your circle, and mark with another stick this spot which will be more towards where the sun rises, and let this second stick be called C, as below you see represented in the drawing.

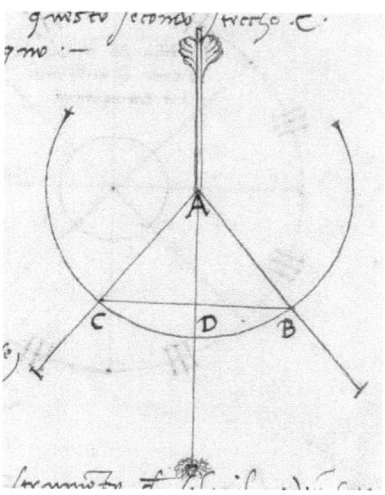

Divide line BC, that is, the distance that there is between one stick and the other, in two equal parts, and let it <the midpoint> be called D, and from point A pull a cord inside the circle to this D. This straight line precisely indicates noon **[in this place]**. With this you can place accurate solar quadrants[79] and all like things.

To know the hour in the night without any other instrument except sight only, do like this. Note, in the evening, when the stars appear, where the north star is, a very well-known star, and keep in mind over which tree or tower or chimney or some like thing it answers to, you being in a given place; and note, of all the stars that move about

---

[78] Translator's note. Gal. 10's 'nona' here means 'noon', and should not be confused with the canonical hour of 'Nones'. Our own English 'noon' comes from the Latin 'nona' or ninth hour, but came to mean midday. See Pierre Souffrin's 2002 translation *Divertissements Mathématiques* [Alberti 2002: 79, note 7].
[79] Translator's note. 'Solar quadrant' is another name for sundial.

7v

tramo[n]tana qualcuna di quelle grandi quale voi possiate facilme[n]te riconoscerla et simile segnate qualch[e] mira in su questa hora la stia sappiate ch[e] in hore • 24 • quella stella ritorna proprio a questo sito dirita a questa mira et tutta hora gira intorno alla tramo[n]tana Adunq[ue] voi la notte quando poi volete nistar l'hora vedete d[e]l tutto el cerchio qua[n]ta parte la cosa verbigratia la fece la quarta parte d[e]l cerchio sono passate sei hore se è il terzo voi il simile p[er] ritrovare la tramo[n]tana si dà certo mezo. Alcuni lo chiamono carro alcuni carrio a similitudine et sono alcune stelle situate come qui vedete la pictura se a vista piglierete p[er] lo cielo una linea quale vada p[er] le dua stelle maggiori ch[e] stan[n]o pari dreto a questa così facta situatione di stelle andando troverete un'altra no[n] piccola stella ne et[iam] molto grande questa prima stella sara dessa et sara scosto da queste dua stelle forse tre[80] volte qua[n]to sia da quelle dua l'una da l'altra chiamono il luogo[81] alcuni quelle stelle le ruote d[e]l carro alcuni la boccha d[e]l corno[82] ma eccho la loro forma:

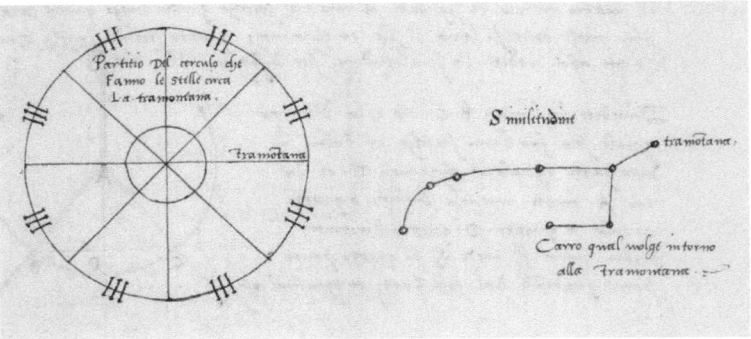

Ma torniamo a qua[n]to mi chiedesti et dicono[83] d[e]le ragioni d[e]l misurar[e] i campi gli scriptori antiqui presertini Columella Savarzarda et altri com[m]ensuratori et Lionardo pisano fra e moderni molto si extese in questa materia et cosa p[ro]lissa et data[84] ma io vi raccolsi le cose più gioconde et anchora sono utili al bisogno

8r.

Non racconto p[er] brevita qua[n]te sieno le forme de campi quadrati et più longo ch[e] largo et più strecto da uno campo ch[e] dal'altro et di tre lati et di piu lati et tondo et parte d'uno tondo et simili tanto dico ch[e] e campi sono co[n] sua lati ò tutti tondi o linee dirite ò parte diritte et parte de archo ò composti di piu[85] altrui come qui vedete la loro varieta seguita.[86]

---

[80] Cfr. Grayson: '...e sarà scosta da queste due dette stelle forse 3 ½ volte...'.

[81] Cfr. Grayson: 'Chiamano **el vulgo** alcuni quelle stelle...'.

[82] Cfr. Grayson: 'Alcuni lo chiamano Carro, alcuni **Corno** a similtudine;...'.

[83] Cfr. Grayson: 'Ma torniamo a quanto mi chiedesti, e **diciamo** delle ragioni...'.

[84] Cfr. Grayson: 'É cosa prolissa e **dotta**'.

[85] Cfr. Grayson: '...e di tre lati, e di **molti** lati,...'.

[86] Cfr. Grayson: '...come qui vedete le loro varietà **segnate**'.

7v

the north star, one of the big ones that you can easily recognise, and in like fashion take note of some marker where it is at this time. You know that in 24 hours that star will come back right to this place, even to this <very> marker, and that it is even now revolving around the north star. Thus, at night, when you wish to judge the hour, look at how much of the whole circle this has transversed. For example, if it has done the quarter part of the circle, six hours have passed; if the third, eight, and so forth. Here is a sure way to find the north star. Some call it the Wagon, some the **[Horn]**,[87] because of its resemblance: these several stars are situated as you see in the drawing. If by sight you take in the heavens a line that goes through the two largest stars that are directly behind each other in the arrangement of stars made like this, continuing on you will find a star that is not small, nor yet very big. This first star will be it <the north star>, whose distance from the two said stars will be perhaps three[88] times as far as the two stars are from one another. Some call those stars in vulgar the Wheels of the Cart, some the Mouth of the Horn. But here is their shape.

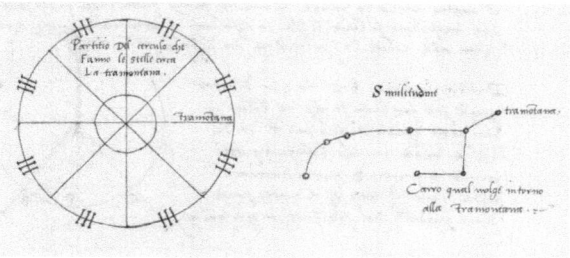

But let us return to what you asked me, and tell about the explanations of how to measure fields. The ancient writers, such as Columella,[89] Savasorda[90] and other surveyors, and Leonardo Pisano[91] among the moderns, went into this subject at length. The thing is quite prolix and known, but I have gathered for you here things that are more amusing, and yet still useful when need arises.

8r.

I will not, for brevity's sake, talk about how many shapes there are of square fields, and <those> longer than wide, or narrower at one end than the other, or of three sides, or several sides, and circular, and part of a circle, and the like. I say that fields are with their sides either all circular, or straight lines, or part straight and part curved, or composed of several curves, as you see their various kinds marked here.

---

[87] Translator's note. This is Ursa Major, called the Big Dipper or the Plough in English.

[88] Translator's note. Grayson gives the distance as 3 1/2 times the distance of the two stars from each other; Souffrin as 4; cf. [Souffrin 2002: 50].

[89] Lucius Junius Moderatus Columella (4 A.D. – ca. 70 A.D.), Roman author of *De Rustica*.

[90] The Latin name for Abraham bar Ḥiyya ha-Nasi (1065 – 1145), Spanish Jewish mathematician, author of an influential treatise on geometry written in Hebrew, and translated into Latin in 1145 as *Liber Embadorum*; see also [Souffrin 2002: 79, note 8].

[91] Leonardo di Pisa, called Fibonacci (ca. 1170 – ca. 1250).

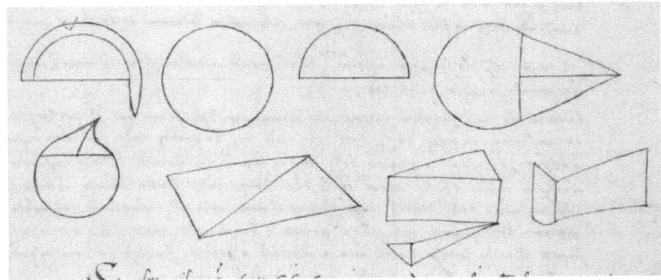

Se volete misurar[e][92] fate così cominciamo da quegli ch[e] hanno e lati tutti diritti et quando e sieno tutti diritti et e cantoni sieno a squadra sara molto facile a intender[e][93] qua[n]ti piedi sia tutto quadrato et farete in questo modo cioè:

Pigliate uno de lati quale volete et notate qua[n]ti piedi sia da l'uno capo a l'altro quando siate da capo volgetevi a l'altra sponde[94] et misuratela forse troverete ch[e] l'uno di questi lati fu x passi et l'altro pur x multiplicate l'uno numero nel'altro chi annovera x p[er] sino a x volte fa cento adunq[ue] sara cento passi quadrato et se efussi passi x p[er] uno lato et passi venti per l'altro sarebbe passi dugento quadrato p[er]ch[è] dieci volti venti fa dugento.[95]

Se e sara di tre facce et uno de sua cantoni sara pure a squadra fate cosi pigliate uno de lati ch[e] termina in sul cantone d[e]l quadro et amoverete qua[n]ti passi eglie poi simile amoverete l'altro quadro

8v

quale simile termina a quel medesimo cantone dello squadro et come facesti di sopra multiplicate l'uno nel'altro et di tutta la soma multiplicata togli la metà et questo sara il vostro campo verbigratia sia l'uno lato passi x et l'altro pur x fara cento le mita è 50 et cosi sara il vostro campo facto a tre canti

Se il campo no[n] sara di queste dua forme dotte[96] et pure sara t[er]minato co[n] linee recte fate cosi habiate una squadra grande et cominciate da uno de lati quale vi par[e] piu apto et secondo ch[e] vi termina la squadra dirizate e fili et cavatene tutti e quadra[n]guli et fate come di sopra multiplicandolo insieme et simile rimane trianguli recti[97] dividendo dove vi pare il luogo più apto et accogliete le somme et star bene.[98]

---

[92] Cfr. Grayson: '**Voi**, se volete misurarle,...'.

[93] Cfr. Grayson: 'Se il campo arà e' lati diritti e i contoni suoi saranno a squadra, e lui sara molto facile ad intendere...'.

[94] Cfr. Grayson: '...quando sete da capo, **continuate e** volgete a **lato** l'altra sponda...'.

[95] Cfr. Grayson: 'Se forse fu [...] dieci per questo e [...] venti per quest'altro [...] , venti volte dieci fa duecento'.

[96] Cfr. Grayson: 'due forme **dette**...'.

[97] Cfr. Grayson: 'E simile, se rimane triangoli, **fate con la squadra vostra di notare gli angoli** retti dividendo dove vi pare...'.

[98] Grayson concludes this paragraph with a further sentence: 'Qui per darvi qualche similtudine posi essempio del modo di dividerli'.

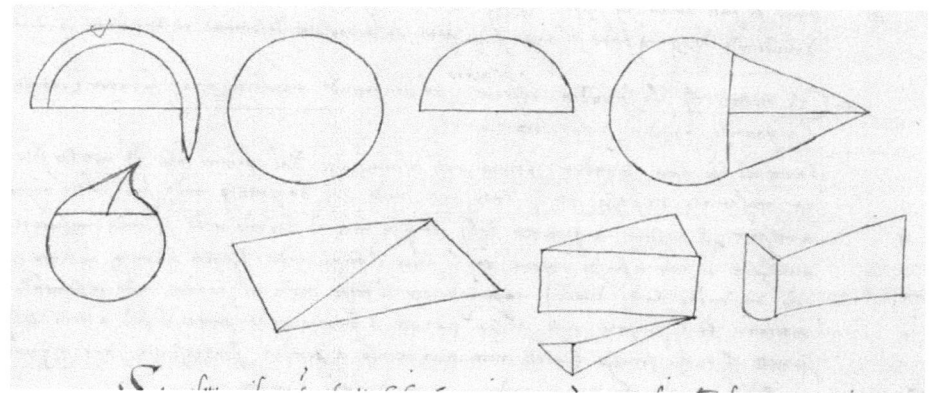

If you want to measure, do like this. Let us begin with those that have sides all straight, and when they are all straight and the corners are right angles, it will be very easy to understand how many square feet there are in all, and you will do it like this.

Take one of the sides, whichever you want, and note how many feet it is from one end to the other. When you are at the end, turn towards the other side and measure it. Perhaps you will find that one of these sides was 10 paces and the other one 10 as well. Multiply one number by the other. He who counts 10 up to 10 times will have one hundred. Thus it will be one hundred square paces. And if it were 10 paces for this side and twenty paces for the other, it would be two hundred[99] square paces because ten times twenty is two hundred.

If it has three sides and one of its corners is still a right angle, do like this. Take one of the sides that ends in the right angle, and count how many paces it is. Then in like fashion count the other side

8v

that likewise ends in the same right angle, and as you did above, multiply one in the other, and of the whole multiplied sum take away half, and this will be your field. For example, let one side be 10 paces, and the other 10 as well, that will make a hundred. The half is 50, and thus will be your field made of three corners.

If the field is neither of these two said shapes and yet it is still bounded by straight lines, do like this. Have a large set square, and beginning from one of these sides that seems most suitable to you, and depending on where the square ends,[100] lay out the lines and divide the whole into quadrangles, and do as above multiplying them together. And likewise, if there remain right triangles, divide the place <in a way> that seems to you to be most apt, and collect the sums, and it will be well.

---

[99] Translator's note. Gal. 10 uses a particularly Tuscan inflection, 'dugento', for the Italian 'duecento', or two hundred.
[100] Translator's note. '...depending on where the square ends': you go along the side until you come to a turn or bend.

Et notate ch[e] la squadra bisogna sia[101] be[n] grande a volerne aver[e] buona certezza la grande squadra meno erra:

farete col filo una squadra optima così cominciate dal primo capo d[e]l vostro filo et misurate tre passi et li fate uno nodo poi da questo nodo piu oltre misurate anchora[102] p[er] insino a quattro passi et qui fate il secondo nodo et indi misurate anchora insino a passi cinque[103] et li fate il terzo nodo harete adunq[ue] in tutto q[ue]sto filo misurato passi duodici raggiugnete il terzo nodo col primo capo et ponetelo in terra et li fichate uno stecho trovate il primo nodo tirate il filo a terra et li ponete il terzo steccho[104] harete uno triangulo a squadra iustissimo sara a squadra quello angulo ch[e] sta al nodo in mezo de passi quattro.[105]

Sono alcuni ch[e] misurono il filo cinque et poi pur cinque et poi sette et sono come uno triangulo[106] questi questi errono perchè e quadrati loro non rispondono a pieno manchavi delle cinque[107] parte l'una et questo basti p[er] e campi ch[e] hanno le linee recte.

Se il campo sara circulare bisogna pigliar[e] la sua largheza et multiplicare[108] tre volte et uno septimo verbigratia se sara largo passi xiiii questo multiplicar[e] p[er] tre et uno septimo ch[e] fa 44 et questa somma sara tutto il suo circuito poi piglate la meta d[e]lla sua largheza quale è 7 et la

9r

metà del suo tondo quale è 22 et multiplicate 7 in 22 somma 154 et questo sara tutto il campo cioè passi 154 ecchovi la figura

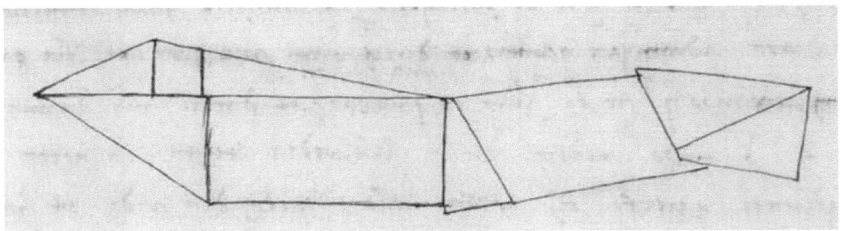

Se il campo sara no[n] rotondo ma circuito da più archi cavatene prima tutti e quadrati ch[e] entrono in tutti e trianguli come diremo[109] di sopra cosi fate et resteranno quelle parte simile a una luna am[m]ezata o scema sella p[ro]prio sara parte qua[n]to uno mezo circulo saprete qua[n]to sarra il tutto p[er] la via di sopra d[e]l circulo et dividetelo p[er] il mezo

---

[101] Cfr. Grayson: 'E notate che la squadra **conviene che** sia ben grande...'.
[102] Cfr. Grayson: 'Poi da questo nodo più **oltre ancora misurate...**'.
[103] Cfr. Grayson: 'indi **ancora seguite e pure misurate, e quando sete in capo di** passi cinque,...'.
[104] Cfr. Grayson: 'e lì ponete **l'altro stecco. Poi ultimo trovate l'altro nodo e simile lì ponete** il terzo stecco'.
[105] Incorrect in both Grayson and Gal. 10; corrected by Grayson. Cfr. Grayson: 'in mezzo de' passi ‹tre e› quattro'.
[106] Cfr. Grayson: '...poi sette, e **fanno come noi** un triangolo'.
[107] Cfr. Grayson: '...màncavi **cinquanta** parti l'una'.
[108] Cfr. Grayson: '...bisogna pigliare la sua larghezza e **multiplicarla** tre volte e un settimo'.
[109] Cfr. Grayson: '...come **dicemmo** di sopra, ...'.

And note that the set square needs to be really large if you want to have a good degree of certainty. The large set square errs less.

Make with a cord an excellent square like this. Start from the first end of your cord and measure three paces, and there make a knot. Then from this knot go further and measure again up to four more paces, and here make the second knot, and then measure further up to five paces and there make the third knot. You will thus have measured in all the cord twelve paces. Join the third knot to the first end and place it on the ground, and stick a stake there. Find the first knot, pull the cord along the ground, and **[put another stake there. Then find the other knot and in like fashion]** there place the third stake. You will have a most correct right triangle which lies between the **[three and]** four paces.

There are some who measure on the cord five and then again five and then seven, and they are like a triangle. These err, because their squares do not completely correspond: one of the **[fifty]** parts is missing. And this is enough for fields that have straight lines.

If the field is circular, it is necessary to take its width and multiply **[it]** three times and a seventh. For example, if it is 14 paces wide, this multiplied by three and a seventh makes 44 paces, and this sum will be its whole circuit.[110] Then take half of its width which is 7, and the

9r

half of its circumference which is 22, and multiply 7 in 22: it sums 154 and this will be its whole area, that is 154 <square> paces. Here is a drawing for you.

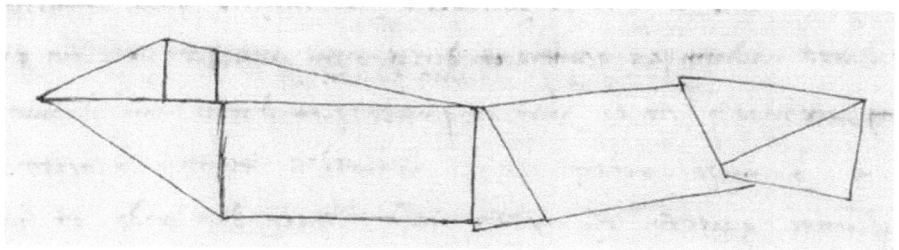

If the field is not circular but surrounded by several arcs, extract first all the squares that go into it, and all the triangles; do as we said above. What remains will be those parts similar to a moon that is half full or less.

---

[110] Translator's note. 'Circuit', i.e., circumference.

se sara parte minor d'uno mezo archulo[111] simile a uno archo gli antiqui feciono una tavola p[er] la quale si misura la corda insino alla steria[112] d[e]larcho et co[n] questa tavola piglavano assai expressa certeza ma sono cose molto intrigate et no[n] apte a questi ludi quali io proposi et qua[n]to attaglia a vostri piaceri basta cavar[e] tutti e quadranguli et tutti e trianguli et ridurli a squadra come dicemo di sopra in questa forma:

9v

Pure se volessi aver[e] qualch[e] principio a comprender[e] la loro ragione, co[n]vie[n]vi divider[e] la corda in due parti et multiplicar luna nel'altre verbigratia sia la corda quattro passi et poi terreti[113] la saetta et multiplicarla nel resto d[e]l diamitro quale se sara uno el resto d[e]l diamitro fa numero qual multiplicato p[er] uno fara quatro sara adunq[ue] quattro et direte uno vie quattro fa quattro quali dua numeri composti cioè uno et quattro mi danno tutto el diamitro ch[e] fa 5 p[er] mezo[114] resta 2 1/2 levatene tutta la saetta cioè dua resta 1/2[115] multiplicate questo ch[e] resta nella meta d[e]la corda et harete tutto el preno di questa parte ch[e] fa 2 3/4[116] questo procede esara meno ch[e] mezzo circulo se esara empierete p[er] questa via quello ch[e] mancha.[117]

Columella pone molto aggiustato certi parti ch[e] a queste misure et questa fara al nostro proposito se la corda d[e]l archo sara piedi xvi la freccia piedi quattro aggiugnete questi dua numeri faranno xx

[111] Cfr. Grayson: '..Se sarà parte e minore che un mezzo **circulo**, ...'.

[112] Cfr. Grayson: '...insino alla **schiena** dell'arco...'.

[113] Cfr. Grayson: 'Verbigrazia, sia la corda quattro passi, **direte due volte due fa quattro**; e poi **torre** la saetta...'.

[114] Cfr. Grayson: '...che sia cinque. **Partite cinque per** 1/2, resta 2 1/2;...'.

[115] Cfr. Grayson: '...levatene tutta la saetta, cioè **1**, resta **1 1/2**;...'.

[116] Cfr. Grayson: '...che fa **3**'.

[117] Cfr. Grayson: 'Questo procede **se** sarà meno che mezzo circulo'.

If it is precisely as much as a half circle, you will know how much the whole will be by the way <described> above regarding the circle, and dividing it by half. If it is part of and less than a half circle, similar to a bow,[118] the ancients made a table to use for measuring the cord at the back of the bow, and with this table they were able to be quite certain; but these are very intricate things and not suitable for these games[119] I am proposing. And as far as suiting your pleasure, it is sufficient to take all the quadrangles and all the triangles and reduce them to squares, as we said above, in this way:

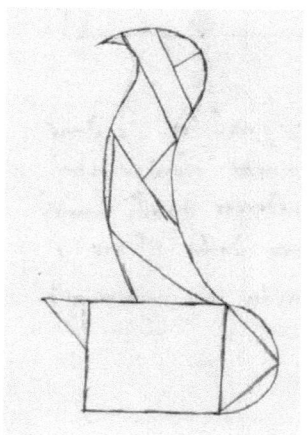 

9v

But if you would still have some principle for understanding their reason, you should divide the cord in two parts and multiply the one by the other. For example, let the cord be four paces, you will say two times two makes four; and then take the arrow and multiple it by the remainder of the diameter, which if it be one, the remainder of the diameter will be the number that multiplied by one makes four. It will thus be four, and you will say that one times four makes four, which two numbers compounded, that is one and four, give me the whole diameter which will be 5, divided by half there remains 2 1/2; take away the whole arrow, that is [1], there remains [1-1/2]; multiply this by what remains in the half of the cord, and you will have all the whole of this part, which makes [3].[120] This works if there is less than a half circle. If it is more, fill in this way for the part that is missing.

Columella sets forth quite correctly certain parts that have this measure, and this suits our purpose. If the cord of the bow is 16 feet, the arrow four feet, add these two numbers, they makes 20.

---

[118] Translator's note. In what follows, Alberti uses a literal bow as a metaphor for a segment of a circle, a literal cord as a metaphor for the chord of a circle, and a literal arrow as the height of the segment, i.e., line segment from the midpoint of the chord to the apex of the arc (where the arrow is placed on a bow). See the commentary for further explanation.

[119] Translator's note. This is the second instance of the word 'ludi'.

[120] Translator's note. Grayson is correct here. 5/2= 2 ½ ; subtracting the arrow (1) leaves 1 ½ , that times 2 equals 3.

an[n]overate questa somma[121] la meta è xl · et d[e]lla lungheza d[e]lla corda la meta è viii quale aggiunta alla meta d[e]lla corda fa xlviii dividete la som[m]a in parte xiiii sara tre et pexo piu[122] quale parte xiiii aggiu[n]ta a 40 fara 54[123] et tanto sara questo archo a similitudine di questo farete questi altri sono queste ragioni molto alte simile molto degne et trate di gran doctrina ma mio proposito è solo qui recitarvi cose iconde adunque lasceremo questa subtilita

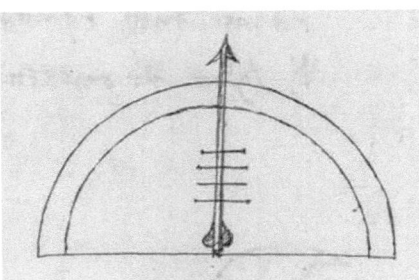

Perch[é] mi chiedesti qual chosa comoda a condurre lacque de fiumi et rivi et simile recitero qualche aptidine[124] rara ma se vorrete veder a pieno et distinta tutta questa materia cioè ch[e] ragione sia di narar[e][125] le vene d[e]le acque [...][126] deducono qual sia il modo di condocti et quali sia l'ordine darvi[127] quale argumento modo et fiumi et rompa eloro empiti in che modo esi vogliono et trasportarsi altrove[128] vederete quelli miei libri di architettura quali io scripsi nel nostro dallo ill[ustrissi]mo vostro fratello mio sig[no]re m[esse]r[129] Lionello et quivi trovarete cose vi dilecteranno assai[130]

10r

Fannosi molti instrume[n]ti p[er] livellar[e] l'acque questo vi piacera pero ch[e] è breve et iustissimo togliete il vostro dardo ò altra cosa ch[e] sia bene diritta e se no[n] havete regolo diritto fate uno archo lungo un passo ò piu et mettetelo in corda et a ciascuno de capi legate uno filo lungo quattro piedi ò più et fate ch[e] e sieno a una lungheza equali et legate e capi di questi dua fili ch[e] pendono insieme et cossi harete facto uno triangulo d[e]l quale da gli lati[131] sono e fili el terzo lato et il dardo ò vero la corda d[e]l vostro arcone

---

121 Translator's note. Cfr. Grayson: 'Annoverate questa somma **quattro volte; sarà ottanta**'.
122 Translator's note. Cfr. Grayson: '...sarà tre e **poco** più;...'.
123 Translator's note. Cfr. Grayson: '...sarà circa a **quarantaquattro**'.
124 Transcriber's note. Difficult to read; Translator's note. Cfr. Grayson: '...qualcuna **attitudine** rara'.
125 Translator's note. Cfr. Grayson: '...cioè che ragione **fu di trovare** le vene dell'acqua,...'.
126 Transcriber's note. Word illegibile. Translator's note. Cfr. Grayson: '...**con che arte si** deducono...'.
127 Translator's note. Cfr. Grayson: '...qual sia l'ordine **de'rivi**,...'.
128 Transcriber's note. Difficult to read because of sign used for 'tr'; Translator's note. Cfr. Grayson: '...transportinsi **altrove**,...'.
129 Transcriber's note. Difficult to read.
130 Translator's note. Cfr. Grayson: 'vi diletteranno [...]'.
131 Translator's note. Cfr. Grayson: '...del quale **due** lati...'.

Count this sum **[four times; it will be eighty.]** The half is forty, and of the length of the cord half is 8; this added to the half of the cord makes 48. Divide the sum in 14 parts, it will be three and a piece more; that 14<th> part added to forty makes **[44]**.[132] That is how much the bow will be. You will treat the others in a similar fashion. These explanations are quite lofty, and likewise very worthy, and treated with great science. But my purpose here is only to tell you about amusing things. Thus we will leave these subtleties.

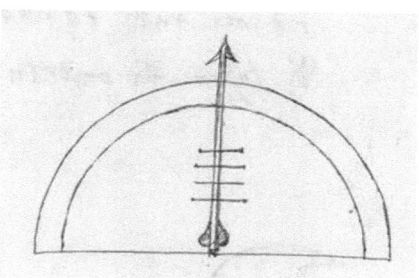

Since you asked me for some easy ways to conduct the waters of rivers and streams and the like, I will recite some rare aptitudes. But if you want to understand fully and distinctly all of this subject, that is, what explanations are given about underground streams, what skills are used to deduce them, how channels are made, what are the arrangements **[of the banks]**, what principle will moderate streams and interrupt their flow, how they are diverted and carried elsewhere, then you have to look at those books of mine on architecture, which I wrote at the request of your most Illustrious brother, my lord, master Leonello, and there you will find things that will quite delight you.

10r

There are many instruments made to level water.[133] This you will like, because it is quick and very precise. Take your spear or anything else that is quite straight, and if you don't have a straight rod, make a bow a pace or more long and string it, and at each of the ends tie a string four feet or more long, and make it so they are of equal length, and tie the ends of these two dangling strings together. Thus you will have made a triangle of which the sides are the strings, the third side is the spear, or the cord of your bow.

---

[132] Translator's note. Grayson's 44 is correct here, and Gal. 10's 54 is incorrect.
[133] Translator's note. To level water is to stop it from flowing by levelling the bottom surface of whatever contains the water.

nel mezo proprio d[e]lla corda d[e]l dardo ò vero d[e]l vostro arco ponete una cera p[er] segno et dove si legano e dua fili insieme legatevi unaltro filo[134] lungo quattro piedi et sieni appicchato uno piombino da laltro capo ch[e] pende et chiamasi questo angulo dove questi tre fili sono annodati insieme · A la prima choccha et capo d[e]l dardo se chiami B la seconda · C la terza in mezo d[e]l dardo si chiami D il piombino · E · come qui vedete disegnato la figura cioè:

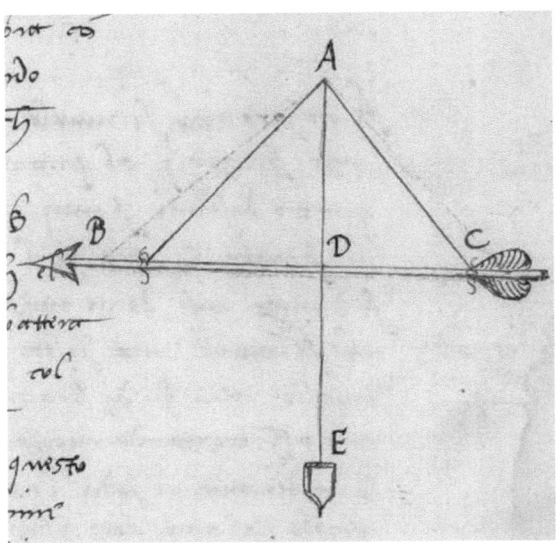

Questo strumento si chiama equilibra co[n] la quale si misura ogni cosa quando lo angulo stara appicchato a cosa ch[e] lo sostenga come si appicha una bi la nera[135] se pesi luno posti al capo · B l'altro al capo · C · saranno equali el filo · A · E · ch[e] pende col piombino battera in su la cera · D adunque voi fate col pr . . . di minuti et pesi[136] ch[e]lla equilibra stia proprio d[e]l pari usasi questo instrumento a piu altre cose maxime a livellar[e] lacque voi mirate p[er] la linea B · C et secondo la sua partita pigliate leltezza[137] d[e]l aqua. M[a] qui molti segniono livellando[138] prima ch[e] no[n] intendono ch[e] la terra sia retonda et vogge in modo ch[e] sempre da qual[e] parte voi siete a livellar par piu alta ch[e] l'altra.

Non mi extendo in dimostrarvi[139] dove sia noto il suo volger[e] et ambito et qua[n]to rispondono inogli[140] migli et gradi d[e]l cielo tanto mi sia p[er]suaso ch[e] in ogni 9000 piedi

---

[134] Translator's note. Cfr. Grayson: '...legatevi un **terzo** filo...'.
[135] Translator's note. Cfr. Grayson: '...**una bilancia**...'.
[136] Translator's note. Cfr. Grayson: 'Adunque voi fate col **porvi e diminuirvi** e' pesi che la equilibra stia proprio **uguale**'.
[137] Transcriber's note. Erroneous transcription for 'l'altezza'; Translator's note. Cfr. Grayson: '...pigliate l'**altezza**...'.
[138] Translator's note. Cfr. Grayson: 'Ma qui molti **s'ingannano** livellando,...'.
[139] Transcribers's note. Word ending unclear.

In the exact middle of the spear or the bow put a bit of wax for a mark, and where the two strings are tied together tie another string four feet long, and there attach a plumb bob from the other end that hangs down, and let this angle, where these three strings are knotted together, be called A. The first corner and end of the spear is called B, the second C, the wax in the middle of the spear D, the plumb bob E, as you see in the figure.

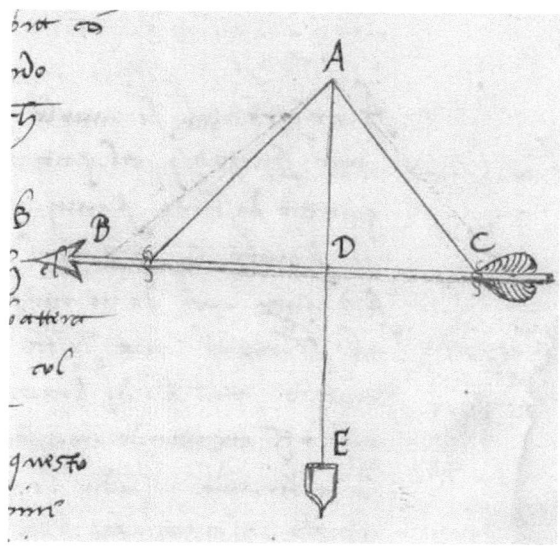

This instrument is called an *equilibra*, with which all things can be measured. When the angle is attached to something that supports it, like <the way> a scale is attached, if the weights placed one at end B, the other at end C, are equal, the string AE that hangs down with the plumb bob will strike exactly on wax D. Thus you put on and diminish the weights[141] so that the *equilibra* is precisely level. This instrument is used for various other things, especially for levelling water. Aim <your line of sight> directly by line BC and according to its division take the height of the water. But here many are fooled when levelling, first because they don't understand that the earth is round and turns so that wherever part you are levelling seems to be higher than the other.

I won't go into to demonstrating to you how its curvature and circumference is known, and how many of your miles correspond to the degrees of the sky. Just be persuaded that in every 9000 feet

---

[140] Transcriber's note. The writing is clear, but the word is not; Translator's note. Cfr. Grayson: '...quanto rispondono **e' vostri** migli....'.
[141] Translator's note. 'put on and diminish the weights': find the level by trial and error.

10v

la terra volge in basso un piè decrinando dalla dirittura di qualunch[e] livella et se volete senza calculo operar[e] livellar[e] di qui in la et segnate le mire alle sue parita[142] et di tutta la differentia pigliate il mezo et questa vi sara apta misurar[e][143]

Anchora no[n] si vuole porre l'occhio troppo presso alla equilibra[144] ma pongasi al quanto discosto p[er] modo ch[e] sotto il veder sieno a uno filo quattro distinti punti cioè la cosa mirate[145] • il punto d[e]le equilibra dua il punto C • tre et il quarto sia le occhio vostro.

Adunq[ue] voi dove misurasti la equalita d[e]l terreno sappiate ch[e] l'acqua nulla no[n] si move ma si sta in collo se l'anno a la china sua almeno p[er] ogni miglio une 1/3 di braccio et questo no[n] vi satisfarebbe se l'anno corressi a dirittina mi p[er]o ch[e] trovando ritoppo di molte soprasta et fermasi se la ripa dove la batte sara ferma et soda l'acqua fa come la palla nel muro quale è mandata a costo al muro pero si parte lungi dal muro se la viene mandata discosto dal muro ella molto discosta donde feri nel muro et fugge ni la intraverso cosi l'impeto de l'acqua se la trova il suo opposito poro oblicho pocho si deduce se ella lo trova molto atraversato ella si deduci assai et batte et rode la ripa co[n]trova onde molti ch[e] no[n] intendono puri riparono indarno alla sua ripa quando doverrieno levar[e] o smmuissare il suo co[n]traposto ò sopra et[146] fare pari une altro traversato onde l'aqua rauvinando co[n]tro al suo contrario imparassi pigliare il corso diritto Anchora l'acqua rode sotto dove la cade et dove la fa alcuno refluo pero ch[e] il peso cadendo il refluo com[m]uove et l'acqua intorbidata correndo porta via, questi principii p[er] ora bastino

---

[142] Translator's note. Cfr. Grayson: '...segnate le mire, **e poi di là in qua e segnate pur le mire** alle sue parità,...'.

[143] Translator's note. Cfr. Grayson: '**atta misura**'.

[144] Translator's note. Cfr. Grayson: 'Ancora **si vuole non por** l'occhio **molto** presso alla equilbra...'.

[145] Translator's note. Cfr. Grayson: '...cioè la cosa mirata, **uno**;...'.

[146] Translator's note. Cfr. Grayson: '...o sopra **sé** fare pari un altro traversato...'.

10v

the earth turns down one foot declining from the straight line of whatever level.[147] And if you want to work without calculations, level from here to there, and mark the targets [and then from there to here and again mark the targets] in equal fashion, and of the whole difference take the half and this will be suitable measuring for you.

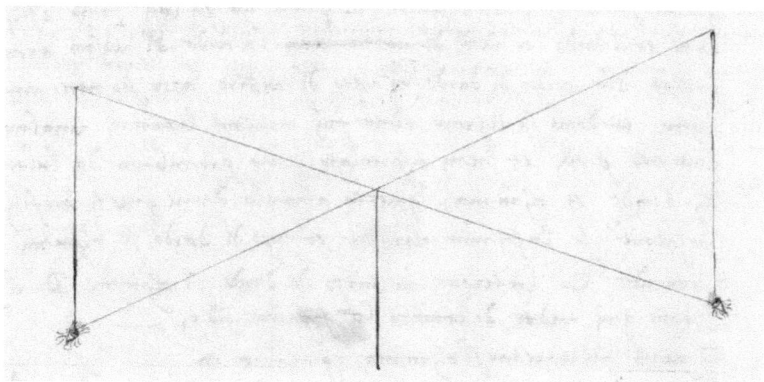

Further, you don't want to put your eye too near to the *equilibra*, but place it at a certain distance so that under the vision are aligned four distinct points, that is, the thing aimed at, [one]; the point of the *equilibra*, two; point C, three; and let the fourth be your eye.

Thus, where you measured the slope of the land, you must know that water does not move at all but stays still if there is not an incline of at least a third of a *braccio* for each mile, and this would not even be enough if it did not run in a straight line, because finding a too high obstacle it would stop.[148] If the bank that it hits is firm and compact, the water will act like a ball against a wall, if thrown <at a> close <angle> to the wall it will move but little away from the wall; if it is thrown far <at a wide angle> from the wall, then it moves back far way from where it hit the wall and flies from <the wall> crosswise. Similarly, the impetus of the water, if it finds its opposite at a slight angle, it bounces back but little; if it finds it quite oblique, then it rebounds a lot, and hits and erodes the opposite bank.

In this way many who don't understand, even while they repair in vain its banks when they should remove or round off the bank opposite, or above it[149] they make equal to it another crossing, where the water running against its opposite, would learn to take the straight course. Further, the water erodes under where it fall and where it makes some reflux, because the weight falling affects the reflux, and the muddied water, flowing, carries it away. These principles are enough for now.

---

[147] Translator's note. 'Whatever level', i.e., in any direction.

[148] Translator's note: This is a very difficult sentence. I believe what he is saying is that even if there is a gentle slope, it is gentle enough so that the water won't have enough impetus to overcome obstacles in its path and will stop. The translation given is literal as per the criteria we set, but a more meaningful translation might be: 'and this wouldn't be good enough for you, if it weren't that the water, running straight, encounters an obstacle that is a number of times greater than it is and stops'.

[149] Translator's note. 'above it': upriver.

11r

Questa equilibra misura ogni peso in questo modo qua[n]to il filo piombinato • A •E • si questa[150] dalla cera • D • tanto quel peso a cui sara piu vicino pesa piu ch[e] l'altro da l'altro capo conoscesi quanto esia cosi qua[n]te volte dal capo d[e]l dardo fino al filo • A • E entra nella parte ch[e] resta d[e]l dardo tante volte luno di questi pesi entra nel altro Verbigratia sia el dardo lungo piedi sei sia dal capo • B uno passo di l[ibre] 4 et dal capo • C un peso di l[ibre] 2 troverete il filo • A • D[151] sarà vicino alle l[ibre] 4 tanto ch[e] quella parte sara di tutta dua et l'altra sara 4 piedi potrei co[n] questa equilibra misurar ogni distantia ogni alteza ogni profondità, ma queste p[er] hora credo bastino ecchovi lo exemplo d[e]l misurar le cose come qui di sotto sara il disegno:

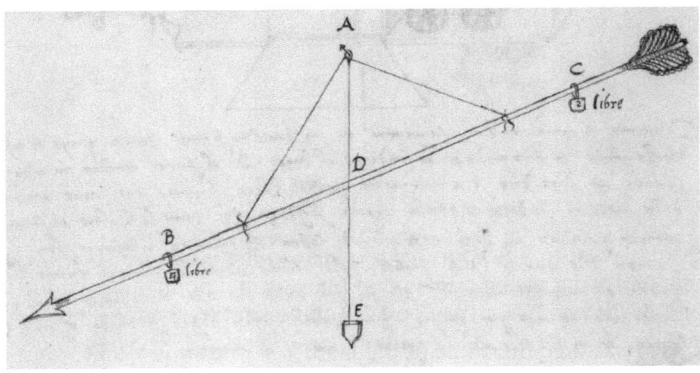

Dapoi faremo[152] menzione de pesi forse sarà a proposito mostrarvi in ch[e] modo si pesi uno sup[er]chio peso come sarebe il carro co buoi et co[n]l suo charchio solo co[n] una stadera ch[e] pesi l[ibre] • L •.[153]

Ordinate uno ponte simile a uno di questi levati et acomodatelo in modo ch[e] le sue chatene ch[e] li stia[154] attachato a uno capo d'una trave lunga quale sia attraversata soppra l'archo d[e]lla porta simile come li adattono i ponti levati et sia da questo luogo della trave dove e posato in sul el suo bilico sopra della porta sino alle catene meno ch[e] del decto bilicho fino da l'altro suo capo ch[e] mene[155] de ento dalla porta et chiamasi el capo delle cathene • A • et il capo drento B • et il bilicho • C • al capo • B • ponete una tagliuola et acomodate il capo d[e]lla fune ch[e] lavorera per questa tagliuola giu entro della porta uno certo naspecto chella charchi

[150] Translator's note. Cfr. Grayson: '...si **scosta** dalla cera D, ...'.
[151] Translator's note. Cfr. Grayson: '...il filo **AE** sarà vicino ...'.
[152] Translator's note. Cfr. Grayson: '**Ma poi che facemmo** menzione de' pesi, ...'.
[153] Translator's note. Cfr. Grayson: '...solo con una statera che **porti** non più che libre cinquanta'.
[154] Translator's note. Cfr. Grayson: '...Ordinate un ponte simile a **questi levatoi**, e accommodatelo in modo **con le sue catene ad alto ch'egli** stia attaccato...'.
[155] Translator's note. Cfr. Grayson: '...suo capo che **vien** dentro alla porta'.

11r

This *equilibra* measures all weights in this way: as much as the weighted string AE is far away from wax D, by that much does that weight to which it is closest weigh more than the other at the other end. You will know how much that is in this way. As many times from the end of the spear to string AE goes into the remaining part of the spear, that many times one of these weights goes into the other. For example: let the spear be six feet long; let there be from end B a weight of 4 pounds, and from end C a weight of 2 pounds; you will find that string [AE] will be 2 feet away from the 4-pound weight, and the other part will be 4 feet. I could with this *equilibra* show you how to measure any distance, any height, any depth, but these [demonstrations] for now I think is enough. Here is the example for you of weighing the things as here below is the drawing.

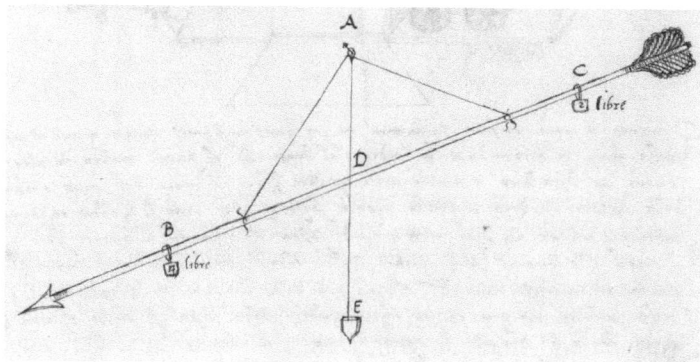

But since we will be making mention of weights, perhaps it suits our purposes to show you how to weigh an extra heavy weight, as would be a cart with the oxen and its load, with only a beam scale that weighs 50 pounds.

Order a bridge <built> similar to one of these drawbridges, and set it up so that its chains are attached at one end to a long beam, which is passed over the arch of the door, similar to how drawbridges are set up. And let it be from this place of the beam where it is placed on its equilibrium point above the door up to the chains, less than from said equilibrium point up to the other end that leads[156] inside the door; and let the end of the chain be called A, and the end inside B, and the equilibrium point C. At end B place a pulley,[157] and set the end of the cable that passes through this pulley, down inside the door in a kind of a spool[158] that takes it up

---

[156] Translator's note: Gal. 10 uses a typical Tuscan term 'menare', to lead or conduct, while Grayson's term 'venire' is more generally Italian.

[157] Translator's note. In his commentary on Alberti's 'tagliuola', Souffrin cites the use of the variants 'tagliolo', 'taglià', and 'taiola', and suggests the French 'moufle', 'pulley-block' or 'pulley', as one would in any case assume from the drawing [Souffrin 2002: 96ff]. He makes the point that a 'palan', i.e., a 'hoist' or 'tackle', is placed between points B and D; see our commentary on Problem 14 for more information.

[158] Tranlator's note. Alberti's term 'naspecto', or 'aspo' is the shaft or shank of a winch. See [Mancini 1917: 201].

11v

et chiamasi • D • questo luogo allato a laltro capo d[e]lla fune ataccharete la vostra stadera accomodata co[n] uno de sua uncini in terra in questa forma et chiamasi questo capo • E come vedete la pictura:

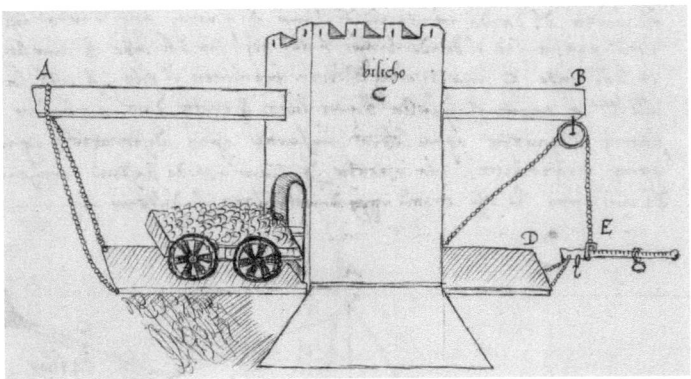

Quando il carro e' buoi saranno in su questo ponte tirate giuso el capo • E • della fine[159] et accomodate la stadera al luogo • D • el ponte andra in alto basta se e va 4 dita dico se annovererete qua[n]te libre il carro porti[160] una oncia della vostra stadera a quella regola peserete poi tutte le altre[161] et sievi ricordo quanto vi dissi teste qui di sopra ch[e] quanta è la parte più lunga d[e]lla trave • A • B • qua[n]te volte ella empie la minore tante l[ibre] porta a numero una l[ibra] ch[e] gli sia posta in capo et la taglia simile qua[n]te volte la fune va giu et su tante volte si parte il peso p[er] modo ch[e] una l[ibra] porta 4 et 6 secondo lo aggirarsi cioè il numero.[162]

Ricordavi[163] ch[e] anchora vi ragionai in ch[e] modo si possa dirizzar una bonbarda sanza veder dove habia adar[e] parmi di no[n] lo preterir più tosto p[er] mostrarvi una praticha d[e]lla vostra equilibra ch[e] p[er] ragionar cosa alcuna[164] d[e]lla degnita et autorita vostra farete così:

fate pesar et notar qua[n]ta polver et ch[e] preta et cochone et zeppe et segnate bene tutto el sito d[e]lla bombarda come la stia posta et addiritta el modo di segnarla certo è questo fate una taccha in su l'orlo fuori della bombarda alla boncha alta in mezo et una lina simile alla corda di qua et di qua a capo et piede fichate stecchi in terra et notate qua[n]to la bonbarda sia discosto da epsi stecchi poi sospendetevi sopra la vostra equilibra et dirizatela a dirittura sopra alle tacche ch[e] sono facte nella bonbarda et notate dove batte il filo piombinato

---

[159] Translator's note. Cfr. Grayson: '...capo E della **fune**,...'.
[160] Transcriber's note. The 'r' is illegible.
[161] Translator's note. Cfr. Grayson: 'Dico che se **una volta** annovererete quante libre del carro porti una oncia della vostra **stateretta**, a quella regola peserete poi **sempre** tutte l'altre'.
[162] Cfr. Grayson: '...secondo il numero dello aggirarsi'.
[163] Translator's note. Cfr. Grayson: '**Ricordami**...'.
[164] Translator's note. Cfr. Grayson: '...per ragionare di cose **aliene** della dignità e autorità vostra'.

11v

and let this place be called D. At the other end of the cable attach your beam scale set up with one of its hooks in the ground in this form, and let this end be called E, as you see in the picture.

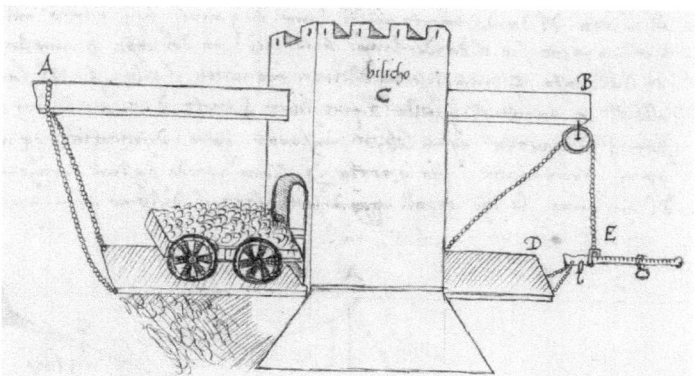

When the cart and the oxen are on this bridge, pull down end E of the cable, and set the beam scale at place D. The bridge will go up. It is enough if it goes four fingers up. I say if you count how many pounds of the cart move an ounce of your beam scale, by that rule you can then weigh all the others. And if you recall what I just said here above, that how many times the longer part of the beam AB contains the lesser, that many pounds corresponds to a single pound that is set at the end; and the pulley similarly, how many times the cable goes up and down, that many times that divides the weight so that a pound carries 4 and 6 depending on the turns, that is the number.

Recall that another time I explained how it is possible to aim a bombard without seeing where the stone has to go.[165] It seems to me that it should not be omitted, more to show you a use for your *equilibra* than to reason about things [alien] to your dignity and authority. Do like this.

Have weighed and noted how much powder and what <size> stone and wood block and wedge, mark well the whole site of the bombard, how it is placed and aimed. The way to mark it surely is this. Make a notch on the outside rim of the bombard at the high mouth in the middle,[166] and another similar one at the head. From here and here at the head and at the foot put sticks in the ground, and note how far away the bombard is from these sticks. Then suspend your *equilibra* above <the bombard>, and align its straight line over the notches that were made in the bombard, and note where the weighted line strikes

---

[165] Translator's note. A bombard is a kind of cannon or mortar used in late medieval times; the projectiles it threw were mainly large stones.
[166] Translator's note. That is, make a notch in the centre of the upper outside rim of the mouth of the bombard.

12r

nella equilibra et qua[n]to ciascuno de sua capi stia lontano et viensu[167] alle dette tocche et p[er] diritura d[e]l capo dove la sta posta mirate il co[n]trario luogo opposito a queste due voi volete tirar[e] et dare la mira d[e]lla vostra equilibra batte ponetevi segno facto questo diesi fuoco alla bonbarda voi vederete dove ella diede et manderetello alto et basso il costiero la seconda volta move[n]do il segno ch[e] voi ponesti la adreto et col segno cosi mosso dirizando la vostra mira equilibra et sotto la equilibra movendo la bonbarda vorrebbesi ch[e] questo segno fussi tanto distante qua[n]to il luogo dove volete dar[e] a trovarlo Adop[er]ate le pratich[e] di sopra eccovi la pictura di questo ch[e] ho detto fino a qui quale raggione molto gioverebbe a chi usassi la balestra ma no[n] mi extendo in ch[e] modo.

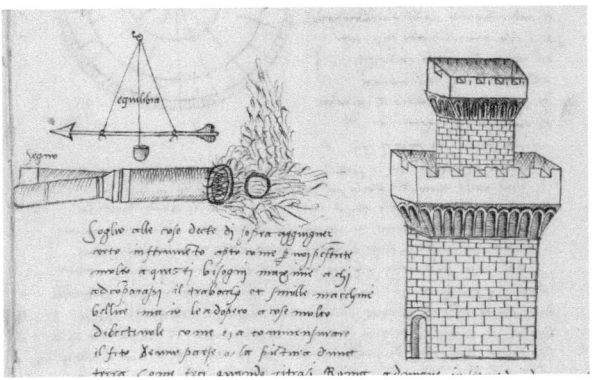

Soglio[168] alle cose decte di sopra aggiugner certo instrume[n]to apto come p[er] voi p[r]eferete[169] molto a questi bisogni maxime a chi adroparassi il trabocco et simille macchine bellice ma io le adopero a cose molto dilectevole come è a commensurare il sito de uno paese ò la pictura d'una terra come feci quando ritrasi Roma adunque in sieme vi daro questa praticha

Misurate il sito et ambito d'una terra et di sue vie et case[170] in questo modo, fate uno circulo in su una tavola largo[171] almeno uno braccio et segnate questo circulo tutto a torno in parte equali[172] quanto volete et qua[n]te più sieno meglio sara purch[é] sieno distinte et nulla confuse Io soglio dividerlo in parte xii iquali tirando diamitro[173] per entro al circulo poi el lembo cioè dintorno tutto parto in parte

[167] Translator's note. Cfr. Grayson: '...stia lontano e **vicino** alle dette **tacche** ...'.

[168] Translator's note. Cfr. Grayson: '**Voglio**...'.

[169] Transcriber's note. Of dubious transcription; Translator's note. Cfr. Grayson: '...per voi **consider(er)ete**,...'.

[170] Translator's note. Cfr. Grayson: '...**cose**...'.

[171] Translator's note. Cfr. Grayson: 'Fate un circulo su una tavola **larga** almeno un braccio,...'. Translator's note: Gal. 10 makes the adjective 'wide' modify the circle, that is, 'circulo ... largo', while Grayson makes 'wide' modify the board, 'tavola larga'.

[172] Translator's note. Cfr. Grayson: '...in parte tutto atorno equali...'.

[173] Translator's note. Cfr. Grayson: '...tirando diametri **tutto** per entro al circulo'.

12r

on the *equilibra*, and how near to or far from each of its ends are to said notches. And along the straight line of the end where it is placed, aim at the contrary place opposite the place where you want to hit, and where the aim of your *equilibra* strikes, put a mark. This done, let the bombard be fired. You will see where it hits, and adjust the high and the low and the slope the second time moving the mark that you had placed behind, and to that mark so moved aligning your *equilibra*, and moving the bombard under the *equilibra*. It should be that this mark is as far away as the place you want to hit. To find it, use the instructions above. Here is the picture of what I have said so far. This explanation would greatly benefit those using a crossbow, but I won't go into how.

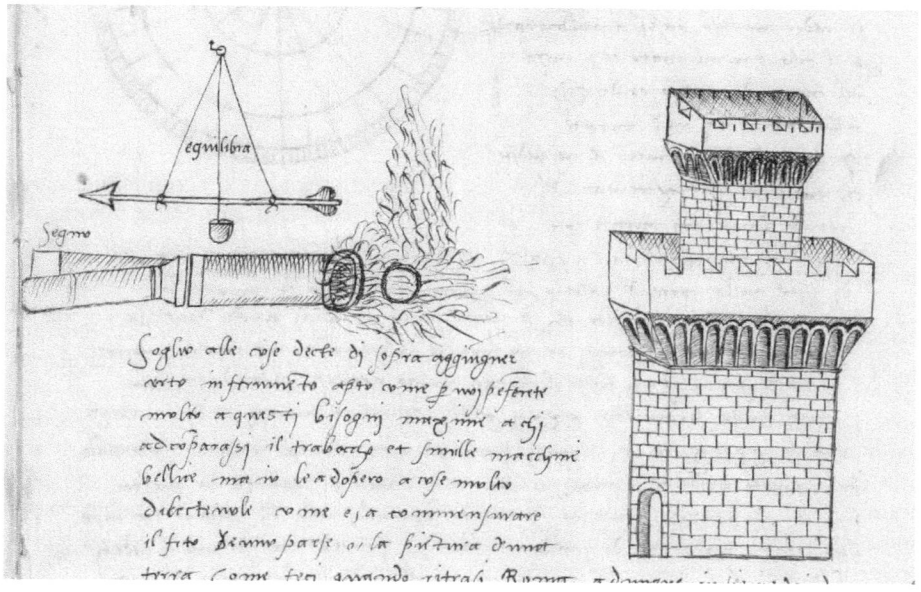

I should add to the things said above a certain instrument very suitable, as you will think, for these requirements, and especially for those using the trebuchet and similar machines of war. But I use it for very pleasurable things, such as how to measure the site of a town, or the picture of a place, as I did when I depicted Rome. Thus together <with the rest> I will explain to you this practice.

Measure the site and perimeter of a place and its roads and houses in this way. Make a circle at least a *braccio* wide on a board, and mark this circle all round in equal parts, as many as you want, and the more the better, since they will be distinct and not at all confused. I usually divide it into 12 equal parts, drawing diameters through the inside of the circle. Then the edge, that is, the entire circumference, I divide into

12v
quara[n]t'otto et queste 48 parte chiamamo[174] gradi è più divido ciascuno di decti gradi in parte 4 et chiamogli minuti et in ciascuno grado scrivo il numero suo simile a questa pittura a pie

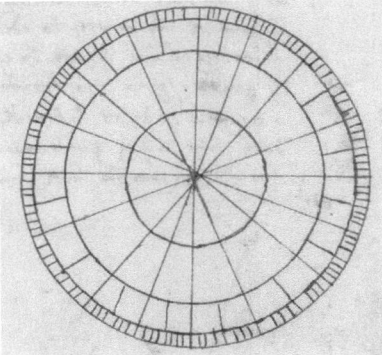

Quando vorrete far la vostra pictura porrete questo strumento in luogo piano et alto dove voi possiate veder multi luoghi d[e]lla terra qual voi volete ritrarre come sono campanili torre et simili et habbiate un filo co[n] uno pio[m]bino et scostatevi da questo strume[n]to dua braccia et guardate[175] a una a una le cose note in modo ch[e] il veder[e] vostro passi a uno riguardo p[er] il filo piombinato et p[er] mezo del centro d[e]l cerchio et dirizisi alla torre qual voi mirate et secondo il numero ch[e] voi vedete ch[e] tagliera[176] alla extremita del circulo dove voi mirate cosi voi fate memoria in su qualch[e] carta di p[er]se verbigratia fingete di esser nella[177] torre d[e]l castello col vostro strume[n]to et mirate la porta lassu et vederete ch[e] il veder[e] passa p[er] venti gradi dove è la divisione dua minuti[178] et no[n] movete lo strome[n]to ma movetevi voi et mirate gli anguli forse il mirar[e] vostro battera sopra dove sono scripti nello strume[n]to xxxii gradi et niuno minuto et piu scrivete anguli xxxii et cosi simili tutti gli altri sanza muover[e] lo strume[n]to facto questo andrete altrove in luogo pure simile et veduto da questo primo et porrete il vostro strume[n]to et statuiretelo ch[e] proprio stia in su la linea medesima di quello numero p[er] il quale voi prima il vedesti al diritto sul vostro strume[n]to cioè ch[e] se da quella torre prima sino a qui una nave havessi a navicar verrebbe per quel medesimo vento segniate ò xx ccii ò vero xxxii et simile qui farete pure il simile come voi facesti al castello noterete d'intorno et farete di tutto memoria in su una altra cartuccia Item piu andrete in altro terzo luogo et pur farete il simile notando tutto et di tutto facendo memoria pongavi la pictura di questo modo[179] qual sara demonstrativa come è detto:

---

[174] Translator's note. Cfr. Grayson: '...parte **chiamo** gradi'.
[175] Cfr. Grayson: '...**mirate**...'.
[176] Cfr. Grayson: 'E secondo il numero che **'l vedere tagliarà** all'estremità del circulo...'.
[177] Cfr. Grayson: '...fingete di essere **sulla** torre del castello...'.
[178] Cfr. Grayson: '...dove è la divisione due minuti. **Scrivete sulla vostra carta: porta di sopra venti gradi e due minuti**'.
[179] Cfr. Grayson: '...modo **che dovete osservare**;...'.

12v

forty-eight parts, and these 48 parts we call degrees. And further divide each of these said degrees into 4 parts, and I call them minutes. At each degree I write its number like in the drawing below.

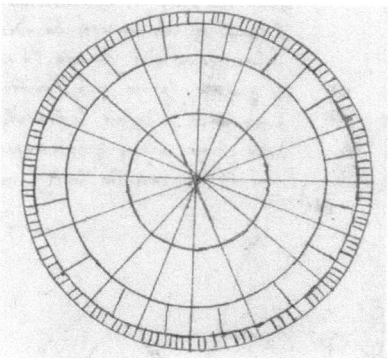

When you want to make your picture, place this instrument in a flat and high place where you can see many places of the land that you want to depict, such as bell towers, towers and the like. And have a string with a plumb, and move away from this instrument two *braccia*, and aim one by one at the things known so that your <line of> sight passes with respect along the weighted string and through the middle of the centre of the circle, and is directed at the tower you are aiming at. And depending on the number that you see that it will cut at the edge of the circle towards where you are aiming, so make a note of this on a piece of paper. For example, pretend you are in the tower of the castle with your instrument and you aim at the gate up there, and see that the <line of> sight passes through twenty degrees where the two-minute division is. **[Write on your paper: upper gate twenty degrees and two minutes.]** And don't move the instrument, but move yourself and aim at the corners. Perhaps your sight will strike where on the instrument it is written 32 degrees and no minutes; again write angles 32. And thus similarly for all the others, without moving the instrument. This done, go to some other place that is similar[180] and visible from the first, and place your instrument, and set it up so that it is precisely on the same line as that number by which you first saw by the straight line on your instrument, that is, if from that first tower up to here a ship had to navigate, it would come by that same wind marked 22 <degrees> 2 <minutes>, or 32 and so forth. And here do in the same way as you did at the castle; look around and make notes of all you see on another piece of paper. Likewise go on to another, third place, and here too do similarly, noting all and of all making notes. I give you a picture of this method which is demonstrative of what has been said.

---

[180] Translator's note. 'Similar', i.e., similarly flat and high.

13r

La linea sta nel suo numero et lo strume[n]to è posto in su quella linea cioè 24[181]

13v

Adunq[ue] farete così chomincerete in su la vostra tavola dove volete fare la pictura et fate uno punto dove vi par apto a la figura di tutta la pictura et questo sia il sito d'uno di queli luoghi donde voi notasti le cose verbigratia sia il castello scrivete qui sopra il facto punto il castello et in su questo punto ponete uno piccolo instrume[n]to di carta largo mezo palmo partito et fatto simile a quel grande dove[182] voi notasti le cose et assettatelo apunto ch[e] il suo centro stia[183] su questo punto et diqui dirizate tutte le vostre linee secondo ch[e] trovate senpio nel vostra memoria simile fate uno secondo ponto[184] nella linea dove vi par toste da voi notata alla tavola qual linea vi nomina uno d[e]lli altri duoi luoghi dove voi mirasti le cose et in su questo punto secondo ponete pure uno simile strume[n]to piccolo di carta et assettatelo ch[e] risponda alla linea quale numero qual nomina su la vostra memoria • Castello • cioè ch[e] l'uno et l'altro strume[n]to sieno una linea insieme rispondenti una a l'altro secondo ch[e] epsi insieme si[185] notate dove voi in su la vostra carta et dove la linea d[e]l primo instrume[n]to vi chiama verbigratia san dom[eni]co si taglia insieme co[n] la linea d[e]l secondo strume[n]to qual pur chiami sancto dom[eni]co quivi fate uno punto et sopra scrivete sancto dom[eni]co

---

[181] Caption not in Grayson.

[182] Translator's note. Cfr. Grayson: '...fatto simile a quello grande **col quale** voi notasti le cose...'.

[183] Translator's note. Cfr. Grayson: '...e assettatelo che ''l suo centro stia **proprio in** su questo punto,...'.

[184] Translator's note. Cfr. Grayson: 'Simile fate un secondo punto **dove vi pare** nella linea **testè** da voi notata alla tavola...'.

[185] Translator's note. Cfr. Grayson: '...che essi insieme **si nominano. E dirizzate ancora quinci tutte le linee al numero loro** notati da voi in sulla vostra carta,...'.

13r

The line is on its number and the instrument is placed on that line, that is, 24

13v

So do like this. Begin on your table where you will make the picture, and make a point where it seems suitable for the figure of the whole picture, and let this be the site of one of those places where you have taken note of the things. For example, let it be the castle; write here above the point made 'the Castle'.[186] And on this point place a small instrument of paper half a palm wide, divided and made similar to that big one where you noted the things, and set it so that its centre is on this point, and here lay out all of your lines according to what you find written in your notes. In a similar way make a second point in the line where you think best just noted by you on the table, which line you call one of the other two places where you aimed at the things, and on this second point place also another similar small paper instrument, and set it so that it corresponds to the line of the number which is called in your notes 'Castle' that is, that both instruments are on a line together corresponding to each other to the line in accordance to what they together [denominate. **And draw again from here all the lines to their number]** noted by you on your paper, and where the line of the first instrument, called, for example Santo Domenico, is cut together with the line of the second instrument, which is also called Santo Domenico, here make a point and write above it 'Santo Domenico'.

---

[186] Translator's note: the names Castle and San Domenico are not in all capital letters in Gal. 10 as they are shown in Grayson, but are set off in some cases by dots. Here I have added the quotation marks for ease of comprehension.

et simili fate di tutte le altre cose et segli accadra ch[e] queste dua linee dette no[n] si taglino bene insieme in modo ch[e] molto sia chiaro il suo angulo ponete uno altro simile instrume[n]to in sul terzo punto donde voi notasti le cose et questo assettate simile agli altri ch[e] fra loro rispondono la loro linea et questo tutto vi manifestera a pieno queste cose a parole et p[er] questo se fanno piu cose come p[er] voi considerete.[187] Con questo diedi modo di ritrovar certo acquedutto antiquo del quale apparivono[188] di molti spirami et erano leni et precluse entro al mo[n]te[189] co[n] q[uest]a via intenderete ch[e] si può notar[e] ogni viaggio et avogimento d'ogni laberinto et d'ogni diserto sanza avoggimento d'alcuno error[e] et co[n] questo potete misurar quanto sia dala torre d[e]l' asinello sino al castello cosi faremo.

Ponete il vostro strumento racconto di sopra diremo p[er] il quale numero si vegga la torre decta et notatelo et poi mirate unaltro luogo alqua[n]to distante da questo dove siete come verbigratia siete da l'uno de capi del corridoro del castello ponete uno certo segno al'altro capo et li miratelo et notate e sua gradi et minuti poi ponete uno certo segno al'altro capo et li miratelo et notate esua gradi et minuti[190] poi ponete il decto strume[n]to in su questo altro capo del coridoio da voi notato et assettatelo come noi diremo ch[e] risponda a uno la sua linea p[er] dirito del corridoio et di q[ui] mirate pur la decta torre et notate in strume[n]to e suoi numeri facto questo habiate in sala ò altrove uno piano uno spatio et

14r

come volessi far[e] la pictura[191] di sopra fafi vostri punti et dirizate la linea co[n] lo instrume[n]to propio come dissi di sopra[192] et dove si taglia segnate in questa forma

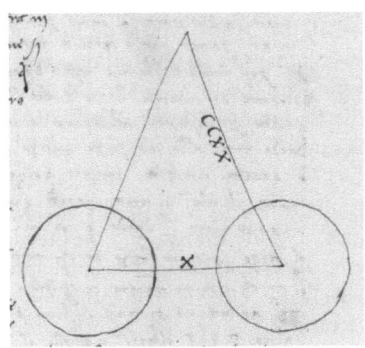

---

[187] Translator's note. Cfr. Grayson: '**Il dimonstrare** queste cose a parole **non è facile, ma la cosa in sé non è difficile, ed è molto delettevole, e con** questo si fanno più cose, come per voi considererete'.

[188] Transcribers's note. Second half of the word unclear; Translator's note. Cfr. Grayson: '...**apparivono**...'.

[189] Translator's note. Cfr. Grayson: '...certo acquedutto antiquo, del quale apparivono **alcuni** spirami ed erono **le vie** precluse entro al monte'.

[190] Translator's note. This sentence is repeated twice in Gal. 10.

[191] Translator's note. Cfr. Grayson: '...e come volessi fare la pittura **detta** di sopra,...'.

[192] Translator's note. Cfr. Grayson: '...proprio come **di sopra dissi**, ...'.

And do the same with all of the other things. If it happens that two of these lines do not cut well together[193] so that their angle is not very clear, place another similar small instrument on the third point where you noted the things, and set this similarly to the others so that their lines correspond to each other, and this will show you everything in full. [**The demonstration of**] these things in words [**is not easy, but the thing in itself is not difficult, and is quite amusing,**] and by means of this many other things can be done, as you can see. By these means I was able to locate a certain ancient aqueduct, of which appeared some remains which were delicate and tunnelled into the mountain.[194] With this way you will understand that you can annotate every journey and the winding of any labyrinth and every desert without involving any errors. And with this you can measure distance very precisely, and if you want to measure how far it is in a straight line[195] from the Torre dell'Asinello to the Castle, we will do like this.

Place your instrument as we said above, according to which number said tower[196] is seen, and make a note, and then aim to another place just as far from that where you are. For example, you are at one of the ends of the corridor of the castle; place a certain mark at the other end, and aim it there, and note its degrees and minutes.

Then place said instrument at this other end of the corridor noted by you, and set it up as we said, so that it corresponds in one of its lines straight down the corridor, and from here aim as well at said Torre, and note on your instrument its numbers.

This done, have in the hall or elsewhere on the floor a space, and

14r

as if you wanted to make the picture above, make your points and lay out the lines with the instrument exactly as I said above, and where they cut it, mark it this way.

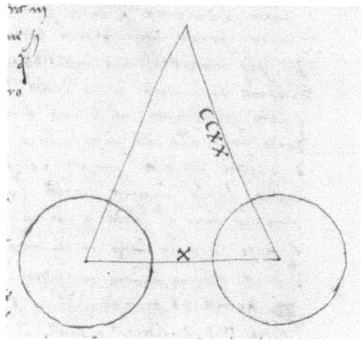

---

[193] Translator's note. 'Do not cut well together', i.e., do not intersect cleanly.

[194] Translator's note. Vasari refers to the project of the aqueduct at Acqua Vergine in his life of Alberti (see [Mancini 1917: 30]). Mancini notes that Vasari seems to be mentioning the restoration of the aqueduct at Acqua Vergine using the geodesic instrument invented by Alberti for the surveying the topology of Rome described in *Descriptio urbis Romae* [Mancini 1917: 31, note 1]. According to Robert Tavernor, no hard evidence links either Prospero or Alberti to this project [Tavernor 1998: 15].

[195] Translator's note: 'in a straight line': as the crow flies.

[196] Translator's note: 'said tower', that is, the Torre dell'Asinello.

Dico ch[e] quante volte lo spatio da luno di questi punti segnato a l'altro entra in[197] queste linee segnato dal punto dove si tagliono tante volte entra lo spatio d' alcuno de capi d[e]l coridoio fino a l'altro nello spatio qual sia dal luogo di quel punto sino a …. [198] vedetelo lì notato fa figura a numeri se da l'uno punto sino a li dove si tagliono le linee sono • d[199] • et direte ch[e] da quello luogo suo d[e]l corridoro fino alla torre sono tante volte qua[n]to è da uno de capi d[e]l corridoio[200] a l'altro[201] et questo vi servira bene a piccole distantie ma alle distantie maggiore bisogna maggior strume[n]to et io vi voglio dar[e] modo ch[e] co[n] tre ciregie misurerete qua[n]to è a dirittura da bolognia a ferrara.

Misurate ogni grande distantia cosi pogniamo caso ch[e] voi vogliate misurare quanto sia a dirittura dal monasterio vostro fino a bolognia andate in su qualch[e] prato grando dove si puo veder[e] bolognia et fichate in terra dua dardi diritti come diremo[202] di sopra ma poneteli distanti luno da laltro mille piedi co[n][203] piu qua[n]to piu vi pare purch[é] l'uno veda l'altro et ciascuno di loro veda bolognia in modo ch[e] tra loro tre cioè bolognia et li dua dardi faccino uno triangulo bene sparto fatto questo cominciate da uno de dardi qual forse sara piu presso verso ferrara et ponetevi co[n] le spalle verso ferrara col viso verso questo dardo et mirate verso il secondo dardo laggiu, adirizando il veder vostro p[er] questo primo dardo et in su questa[204] linea ch[e] fara in terra il vostro veder[e] di lungi al dardo xx piedi ponete uno segno et se e piaccessi a voi sia una coreggia poi volgetevi al viso verso bolognia et mirate p[er] dirittura di questo medesimo dardo et in terra simile nella linea qual fara il vostro veder di lungi xxx piedi ponete una rosa co[n][205] quello vi piace harete adunque notato in . . .[206] uno triangulo d[e]l quale uno angulo

14v

sara verso ferrara cioè il dardo verso il mar[e] sarà una ciregia, verso bolognia sara una rosa, chiamasi adunq[ue] il dardo qui • A • la ciregia • B • la rosa • C[207] • et notate bene queste misure apunto, fatto questo ite al secondo dardo[208] mirate a dirittura al dardo primo et p[er] questa dirittura qual fa il vostro mirar[e] ponete una ciregia presso a questo dardo

---

[197] Translator's note. Cfr. Grayson: '…entra in **una di** queste linee…'.
[198] Translator's note. Cfr. Grayson: '…sino **all'Asinello**'.
[199] Transcriber's note. Illegible symbol; Translator's note. Cfr. Grayson: '…sono **once duecentventi**,…'.
[200] Transcriber's note. Partially illegible due to a correction made.
[201] Translator's note. Cfr. Grayson: 'Se dall'uno punto **all'altro è once dieci, e da questo punto** sino a lì dove si tagliano le linee sono **once duecentventi**, direte che da quello luoto suo del corridoio sino alla Torre **dell'Asinello** sono **ventidue** volte quanto è da uno de' capi del corridoio all'altro'.
[202] Translator's note. Cfr. Grayson: '…come **dicemmo** di sopra,…'.
[203] Translator's note. Cfr. Grayson: '…mille piedi **o** più quanto vi pare,…'.
[204] Translator's note. Cfr. Grayson: '…e su **quella** linea…'.
[205] Translator's note. Cfr. Grayson: '…ponete una rosa **o** quello vi piace'.
[206] Translator's note. Cfr. Grayson: 'Arete dunque notato in **terra** un triangulo,…'.
[207] Translator's note. Cfr. Grayson: '…la rosa C. **Misurate quanto sia da B ad A, e quanto da A a C, e da C a B,** e notate bene queste misure appunto'.
[208] Translator's note. Cfr. Grayson: '…al secondo dardo, **e volgete il viso verso Ferrara, e scostatevi venticinque piedi, e per questo secondo dardo** mirate a dirittura il dardo primo, …'.

I say that the number of times the space from one of these marked points to the other goes into [one of] these lines marked by the point where they cut, that many times the space from one of the ends of the corridor to the other goes into the space that is from that point to [the Torre dell' Asinello]. See the figure noted with numbers. If from one point [to the other is ten *once*,[209] and from this point] to where it cuts the line are [two hundred twenty *once*], you will say that from that place of yours in the corridor up to the torre [dell'Asinello] are [twenty-two] times the distance from one end of the corridor to the other end.

And this will serve you well for small distances, but for larger distances you will need a larger instrument. And I want to give a way that with three cherries you can measure how far it is in a straight line from Bologna to Ferrara.

Measure any large distance like this. Let us suppose that you want to measure how long it is in a straight line[210] from your monastery up to Bologna. Go up to some large field where Bologna can be seen, and stick two spears straight into the earth, as we said above, but place them one from the other at a distance of a thousand feet and as much more as you wish, as long as one sees the other and each sees Bologna, so that between the three of them, that is, Bologna and the two spears, they make a well partitioned triangle. This being done, begin from one of the spears, the one perhaps that is closest to Ferrara, and place yourself with your back towards Ferrara and face towards this spear, and aim <your sight> towards the second spear down there, directing your sight through the first spear here; and on this line that your sight will make on the ground, twenty feet away from the spear make a mark, and if you like, let it be a cherry. Then turn with your face towards Bologna, and aim <your sight> along the straight line of this same spear, and on the ground in a similar fashion on the line that your sight will make, thirty feet away place a rose with whatever else you like. You will have thus noted on the [ground] a triangle, of which one angle

14v

towards Ferrara will be the spear, towards the sea it will be a cherry, and towards Bologna it will be a rose. Let the spear here be called A, the cherry B, the rose C [.Measure how much it is from B to A, and how much from A to C, and from C to B], and carefully note these precise measures. This done, move to the second spear, [and turn your face towards Ferrara, and move away twenty-five feet, and along this second spear] aim in a straight line at the first spear, and along this straight line made by your <line of> sight, place a cherry as near to this spear

---

[209] Translator's note: In this case an *oncia*, a unit of length, is one-twelfth of a *braccio*; later *oncia* is used as a unit of weight (see fol. 15v, pp. 66–67 in this present volume).

[210] Translator's note. 'in a straight line', that is, as the crow flies.

proprio come[211] stava • B presso a A • poi volgete il viso verso Bolognia, et p[er] la dirittura di questo dardo mirate Bolognia et in terra in su quella linea ponete una rosa distante dal dardo primo qua[n]to fu distante nel primo triangulo • C • da • A[212] • et torrete une filo da questo dardo fino alla rosa fatto questo tornate dove ponesti la ciregia et p[er] dirittura di questa ciregia mirate bolognia et notate bene dove questo mirare toste taglia il filo[213] in terra harete qui notato une altro triangulo quale uno angulo fara il dardo • chiamasi • D • le altro sara la ciregia et chiamasi • E • el terzo sara lo stecho et chiamasi • F • et p[er] meglio exprimer eccovi la simile pictura

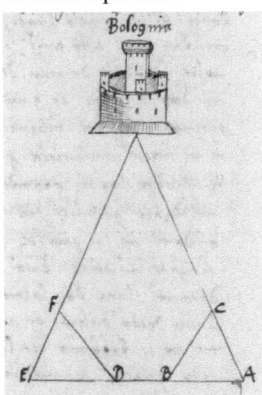

Dico ch[e] qui vi co[n]viene considerare ch[e] voi harete tre trianguli luno •A•B•C• l'altro •D•E•F• il terzo è quello il quale li anguli suoi sono luno bolognia l'altro il dardo • A l'altro la ciregia • E • misurate qua[n]te volte entra la linea • E • D • nella linea • E • F • nel suo piccolo triangulo tante volte • E • A enterra in tutta la linea • F[214] • p[er] insino a bolognia nel suo grande triangolo per meglio exprimer eccovi[215] lo exemplo a numeri sia • D • E • dieci piedi et sia la linea[216] • E • F • xxxx piedi dico ch[e] come esi entra in xxxx iiii[217] volte cosi la linea et spatio • E • A • enterra volte xxxx nella linea et spatio fra • E • et bolognia[218] et se • E • D enterra xxx volte in • E • F da qui[219] dove voi operate sino a bolognia sara xxx volte qua[n]to sia da • A • in fino a E •

---

[211] Translator's note. Cfr. Grayson: '...ponete una ciriegia presso a questo dardo **primo quanto** stava B presso ad A'.

[212] Translator's note. Cfr. Grayson: '...una rosa distante dal dardo **proprio** quanto fu nel primo triangulo **distante** C da A,...'.

[213] Translator's note. Cfr. Grayson: '...e notate bene dove questo mirare **testé batte in terra e** taglia il filo **posto e tirato fra 'l dardo e la rosa, e qui ponete una bacchetta'**.

[214] Translator's note. Cfr. Grayson: '...tutta la linea **E**...'.

[215] Translator's note. Cfr. Grayson: '...eccovi **del tutto** l'essemplo a numeri'.

[216] Translator's note. Cfr. Grayson: '...e sia [**...**] EF quaranta piedi'.

[217] Transcriber's note. Word crossed out here, and 'iiii' written above.

[218] Translator's note. Cfr. Grayson: 'Dico che come **dieci** entra in **quaranta quattro** volte, così la linea e spazio EA enterrà volte **quattro** nella linea e spazio fra E e Bologna;...'. Grayson's

[219] Translator's note. Cfr. Grayson: '...**qua**...'.

exactly as much as B was near to A. Then turn your face towards Bologna, and along the straight line of this spear aim at Bologna, and on the ground on that line place a rose that is as far from the first spear as C was from A in the first triangle, and take a cord from this spear up to the rose.[220] This done, go back to where you placed the cherry, and along the straight line of this cherry aim at Bologna, and mark well where this aim <strikes the ground> and cuts the cord placed and stretched between the spear and the rose, and here place a stick. You will have noted here another triangle, of which one angle will be the spear, let it be called D, the other the cherry, let it be called E, and the third will be the stick, let it be called F. And to express this better, here is its likeness in a picture.

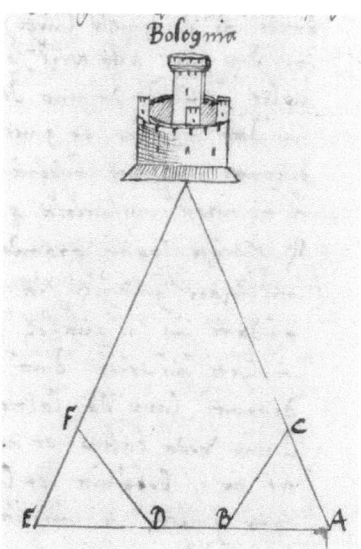

I say that here it behoves you to consider that you have made three triangles, the one, ABC, the other DEF, the third that in which the angles are the one Bologna, the other the spear, the other the cherry E. Measure how many times line ED goes into line EF in your small triangle, how many times EA goes into the whole line **[E]** all the way to Bologna in your large triangle. To express it better, here is the example in numbers for you. Let DE be ten feet, and let the line EF be 40 feet. I say that since 10 goes into 40 4 times, so the line and space EA will go 4 times into the line and space between E and Bologna; and if ED goes 30 times into EF, from here where you are working up to Bologna will be 30 times as much as it is from A up to E.

---

[220] Translator's note. This sentence is notoriously problematic; see our commentary for an extended discussion.

ma p[er]ch[é] no[n] si[221] possono sempre veder a occhio le distantie et giova saper propio qua[n]to la cosa sia distante vi daro modo da misurar[e] qua[n]to sia da ferrara sino a milano et farete cosi me[n]tre dormirete et giacerete et di tanta misura harete certeza fino a uno braccio[222]

Habbiate uno carro quanto maggior di ambito sono le ruote meglio

15r

fia in sul moto[223] grosso d[e]lla ruota in quale stanno confitti erazi et nel quale entro p[er]tusato passa quello ch[e] altrui chiama a xis cavate una fossetta no[n] maggior ne piu profonda se no[n] qua[n]to riceva una sola ballotta et fate una casa col suo p[er]fuso sopra il vostro moto in modo ch[e] nessuna balotta eschi se no[n] quando volgiando si la ruota una sola ne entri nella sua fossetta empiete questa cassetta di palloctole et fatto fattevi dove quando volgendosi[224] sia credo p[er] vostro ingegno intendete come p[er] le ballotte[225] vi saranno note le volte della ruota e a voi sia noto qua[n]to volge la ruota co[n]terete adunque tante ballotte tante ruote et tante volte tante braccia eccovi la pictura:

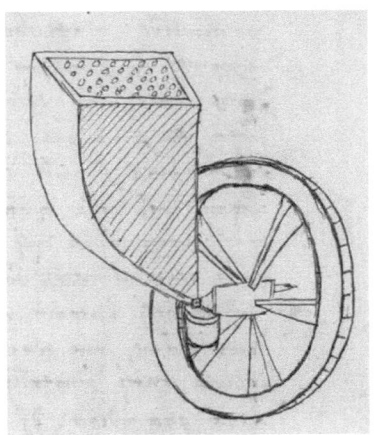

Questo medesimo modo si puo adoperar a conosser la via p[er] mar facendo la ruota in sca[m]bio de razi sieno pale simile a quelle del mulino et appenderla al lato della nave del resto farmi il simile ch[e] io dissi di sopra una fossicella nel fuso quale entrasi nella nave ma voglio darvi certo modo a saper[226] qua[n]to la vostra fusta vadia p[er] hora a qualunch[e] ve[n]to la muova fate così

---

[221] Transcriber's note. Difficult to interpret.
[222] Translator's note. Cfr. Grayson: '...da Ferrara sino a Milano **giacendo e dormendo**, e **in** tanta misura arete certezza per insino ad un braccio. **Farete così**'.
[223] Translator's note. Cfr. Grayson: '**motto**'; modern Italian 'mozzo', wheel or hub.
[224] Cfr. Grayson: '...e **sotto** fatevi dove, quando volgendosi **la ruota lasci la pallotta riceuta nel pertuso fatto sotto, sia ricolta, o sacco, o che si** sia'.
[225] Translator's note. Cfr. Grayson: '...intendete come **secondo il numero delle pallotte cadute** vi saranno note...'.
[226] Translator's note. Cfr. Grayson: '...modo raro a **conoscere** quanto...'.

But since it is not always possible to see the distances by eye, and it is good to know precisely how distant the thing is, I will give you a way to measure how far it is from Ferrara to Milan while you are sleeping and lying,[227] and of that measure you can be certain to within a *braccio*. Have a cart; the larger the circumference of the wheels, the better.

15r

In the large hub of the wheel, in which the spokes are fixed, and in the perforation of which passes what the Latins call the *axis*, make a little hole neither larger nor deeper than into which a single little pellet can fit. And make a box with its opening above your hub of the cart, in such a way that no pellet comes out unless when, as the wheel turns, only one enters into the hole. Fill this box with pellets, and underneath make a place **[where the pellets, released through the hole in the bottom as the wheel turns, are collected, a sack or]** whatever. I believe you have wits enough to understand that by the pellets, you will know the number of rolls and you know how far <one> roll <of the wheel> is. Count therefore how many pellets, how many rolls, and that many times, that many *braccia*. Here is the picture.

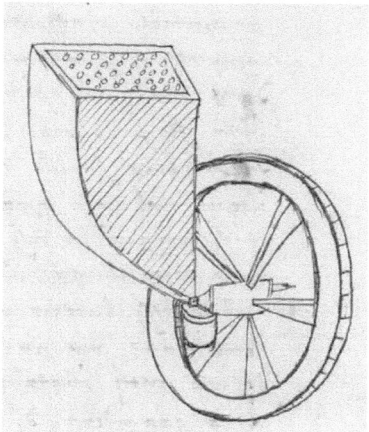

This same method can be used to know the distance at sea, making the wheel so that in place of spokes are vanes like those of mills, and hanging them on the side of the ship; do the rest in like manner as I described above, a little hole in the shank which goes into the ship. But I want to give you a certain rare method for knowing how fast your fusta[228] will go per hour in whatever wind blows. Do like this.

---

[227] Translator's note. 'sleeping and lying': doing nothing at all (in other words, the mechanism described below does the work for you).

[228] Translator's note. A fusta is a narrow, light and fast ship, also called a foist or a galliot.

A conoscer qua[n]to navichi una vela ponete il vostro pennello facto no[n] di piume ma di legno fitto nella sua Astucla[229] et habiate una assicella sottile qua[n]to uno gavioro[230] lungo uno pie larga quattro ditta apichate co[n] dua quercetti giu basso alla corda[231] del pennello ultimo in modo ch[e] la simuovia[232] no[n] qua et la verso ma destra et sinistra qual fa il suo pennello et su et giu come fanno gli usci ma su et giù come fanno le casse quando le aprite et serrate et fievi una parte d'uno arco qual penda in giu attachato in modo ch[e] quando questa assicella stara più alta ò più bassa voi possiate ivi nel decto archo tutto segnare[233] et più chiareza[234] vostra eccovi la similitudine di questa asse pennello et arco: [235]

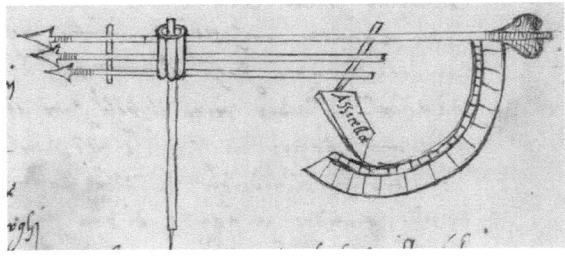

Questo no[n] bisognia persuadervi ch[e] qua[n]do no[n] tireran[n]o ve[n]ti questa assicella pe[n]dera giu a dirittura[236] qua[n]do sara poco ve[n]to questa poco si alzera et qua[n]do sarà forte ella stara sollevata assai co[n]vienvi haver adunque notato et bene co[n]siderato altrorove[237] a luoghi noti a voi qua[n]to la vostra fusta corre per hora et p[er] tanto ch[e] lassicella salzi

15v

a questo ò a queste altro segnio co[n] queste vele tanto alte ch[e] cosi adiritte co[n] questo charicho co[n] tanti trinati[238] in acqua è simile a questi segni et notatione poterli ch[e] vi sieno bene certissimi et presenti adunq[ue] navicando[239] la vostra fusta qua[n]te hor la corse p[er] il vento d[a]l tale segnio ò laltre circunstantie a voi note et cosi harete certa notitia d[e]l vostro navigio et no[n] co[n]verra arbitrar p[er] altre coniecture le miglia come oggi fanno e marinari.

---

[229] Transcribers's note. Difficult to interpret; Translator's note. Cfr. Grayson: '...**astola**...'.

[230] Transcribers's note. Difficult to interpret; Translator's note. Cfr. Grayson: '...sottile quanto un **cuoio**,...'.

[231] Translator's note. Cfr. Grayson: '...giù basso alla **coda** del pennello ultima,...'.

[232] Transcriber's note. Partially illegible; Translator's note. Cfr. Grayson: '... che ella si muova...'.

[233] Translator's note. Cfr. Grayson: '...tutto segnare **e annotare**'.

[234] Translator's note. Cfr. Grayson: '...et **per** più chiarezza...'.

[235] Translator's note. Cfr. Grayson: '...di questo pennello e asse e arco'.

[236] Translator's note. Cfr. Grayson: '...a dirittura, **e** quando sarà poco vento,...'.

[237] Transcriber's note. Probable errorroneous transcription for 'altrove'. Translator's note. Cfr. Grayson: '... e ben **conosciuto altrove** a ...'.

[238] Translator's note. Cfr. Grayson: '...con tanti **timoni** in acqua...'.

[239] Translator's note. Cfr. Grayson: '...Adunque navicando **porrete mente quante ore corse** la vostra fusta pel vento del tal segno, ...'.

To know how much a sailboat sails, place your pennant, not made of feathers but of wood, driven into its staff, and have a little board as thin as leather, a foot long, and four fingers wide.

Attach it with two little oaken pegs down below at the last end of the pennant, such that it doesn't move here and here to the left or the right, as your pennant does and as doors do, but up and down like chests do when you open or close them; and let there be a part of an arc which hangs down attached so that when this little board is higher or lower, you can there mark everything on said arc. And [for] your greater clarity here is a likeness of this board, pennant and arc.[240]

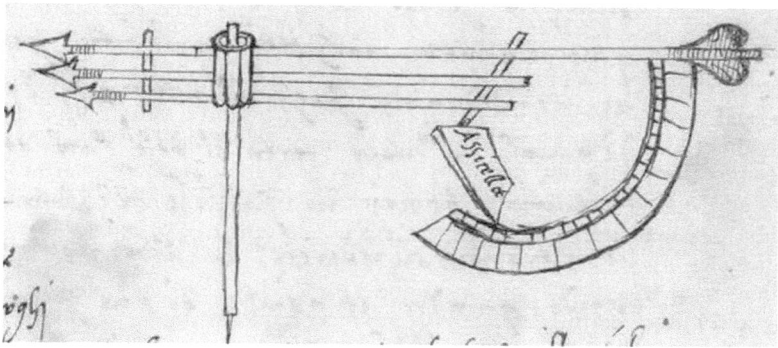

Here there is no need to persuade you that when no winds blow, this little plank will hang straight down; when there is a little wind, this will lift up slightly, and when the wind is strong, it will be quite lifted. It behoves you in any case to have noted and well ascertained elsewhere in places known to you how fast your fusta goes per hour, and with how much wind the little plank goes up

15v

to this or this other mark, with these sails at this height it goes straight as this, with this load, with how many rudders in the water, and the like; and these marks and notations make them so that they are extremely precise and clear. Thus when sailing [keep in mind] how many hours your fusta has run with the wind at this mark, with other circumstances that you know, and so you will have certain information about your navigation, and you need not estimate the miles by means of other conjectures like sailors do today.

---

[240] Translator's note. Alberti is describing an anemometer, a device for measuring wind speed, and this is apparently its first description in writing. Alberti is credited with its invention; a second instance is that of Robert Hooke (seventeenth century).

Anchora piglierete piacer di questo ch[e] gli antiqui scripsono come Ierone principe di seracusa fece certa op[er]a d'oro di molto peso et di grande magistero quale facta rispondeva nella bilancia al peso d[e]l oro quale egli haveva dato a maistri ma intese ch[e] e maistri artifici d[e]l opera lo havevano inganato et ch[e] e no[n] era tutto el lavoro d'oro ma era misto d'argento irato Ieron[e] no[n] voleva pero guastar il lavoro ma voleva certificarsi com[m]isse archimede matematico questa causa Archimede huomo sottilissimo sanza muover ò guastar nulla tutto vidde manifesto ecco il modo

fecie due masse d'uno medesimo peso qua[n]to fu l'opera facta da e maistri et di queste due masse luna fu puro oro, l'altra puro argento posselle nella acqua ne vasi a una gra[n]dezza et natura²⁴¹ simili et pieni a un modo et vide ch[e] differentia restava di questa acqua nel vaso quando ponendovi questa massa l'acqua trabocchava fuori et si versava et cosi posevi poi l'opera et proportionando e pesi loro insieme trovò certo il vero, in tutto el lavoro fu ingegnio molto acuto.

Qua[n]to pesi l'acqua a proportione d[e]l oro no[n] scripsono li antiqui pero ch[e] l'acque sono varie ma trato²⁴² bene scripto qua[n]to a proportione d[e]lla cera pura pesino tutti i metalli et diro ch[e] uno dado ò palla ò qual forma si sia di certa grandezza di cera ch[e] pesi una oncia questa medesima sendo di rame puro pesera o[ncie] viii et danari uno se sarà stagnio peserà o[ncie] 12 se sarà piombo peserà o[ncie] 18 d[enari] 6²⁴³ se sara oro pesera libre una o[ncie] 7 d[enari] 9 di qui si puo facile conoscer qua[n]to pesi più l'oro nella acqua ch[e] lo argento²⁴⁴ et la ragione è evide[n]te sì ch[e] qualunch[e] capo²⁴⁵ essendo pari a misura co[n] lacqua et in se pesi meno questo stia tanto sollevato et a galla qua[n]to il suo peso sara minor et stara pari immerso nel acqua quanto pari tanta qua[n]tita d[e]l'acqua sara di peso pari a lui et a queli cap²⁴⁶i ch[e] in se pesono piu ch[e] lacqua staranno sotto et qua[n]to piu peseranno tanto piu veloci discenderanno et meno occuperanno delacqua sendo tutti d'una figura et forma co[n] questa ragione mostrai qui a questi architetti quanto pesi certa colonna di quale epsi tra loro contendevano²⁴⁷ presi alami²⁴⁸ pezi di simile pietra et alcuni di marmo de quali io ho noto certo il peso suo et posili nel'acqua et compresi la loro differentia potrei in simile cose molto extendermi ma questi p[er] hora bastino

---

²⁴¹ Translator's note. Cfr. Grayson: '...una grandezza e **una forma** simili...'.

²⁴² Translator's note. Cfr. Grayson: 'Ma **truovo** bene scritto...'.

²⁴³ Translator's note. Cfr. Grayson: '...sendo di rame puro peserà oncie otto e denari **sedici, e se sarà di rame ciprino, peserà oncie otto e denaio** uno; se sarà stagno, peserà oncie dodici; se sarà piombo, peserà **una libra e denari sei**;...'.

²⁴⁴ Translator's note. Cfr. Grayson: 'Di qui si può facile comprendere per che cagione l'oro pesi nell'acqua più che l'ariento,...'.

²⁴⁵ Translator's note. Cfr. Grayson: '...qualunque **corpo** essendo...'.

²⁴⁶ Translator's note. Cfr. Grayson: 'E quelli **corpi**...'.

²⁴⁷ Translator's note. Cfr. Grayson: '...contendevano fra loro'.

²⁴⁸ Transcriber's note. Possible erroneous transcription for 'alcuni'; Translator's note. Cfr. Grayson: '...Presi **alcuni** pezzi di simile pietra...'.

Further, you will enjoy hearing what the ancients had to say about how Hieron, prince of Syracuse,[249] had commissioned a certain work of gold of great weight and made with great mastery <and> which was supposed to correspond on the scale to the weight of the gold that he had given to the master artists. But he understood that the master artists of the work had fooled him and the work wasn't all of gold but was mixed with silver. Angry, Hiero did not, however, want to ruin the work, though he wanted proof. He committed the mathematician Archimedes to this cause. Archimedes, a most cunning man, without moving or breaking anything, made everything manifest in this way.

He made two masses of the same weight as the work made by the master artists, and of these two masses one was pure gold, the other pure silver. He placed them in water in vessels of like size and nature and filled to the same level, and saw what difference <there was> in this water in the vessels, when each mass was put in it and the water overflowed and spilled <out>. And so he then placed the work, and proportioning their weights together he found for sure the truth in the work as a whole. It was a very acute contrivance.

How much water weighs in proportion to gold the ancients didn't write, because waters are of various kinds. But I find well written that concerning the proportion of pure wax that weighs all metals. And they say that a cube or ball or whatever other form of a given size of wax weighing an *oncia*, this same <shape> being of pure copper will weigh 8 *once* and **[sixteen *denari* and if it is of Cypriot copper, it will weigh eight *once* and]** one *denaro*; if it is of tin, it will weigh 12 *once*; if it is of lead, it will weigh 18 *once* and 6 *denari*; if it is of gold, it will weigh one *libra*, 7 *once* and 9 *denari*. From this it is easy to know how much more gold weighs in water than silver, and the explanation is evident. So then any item being of equal measure with the water and in itself weighing less, this will be lifted up and <kept> afloat by as much as its weight is less, and it will stay likewise immersed in the water by as much as the quantity of water will be of a weight equal to it. And those items that in themselves weigh more than the water, will stay under, and the more they weigh, the faster they will sink and less of the water they will occupy, being all of the same shape and form. With this explanation I recently demonstrated to these architects here how much a certain column about which they were arguing weighed. I took some pieces of like stone and some of marble whose weight I had precisely noted, and I put them in water and noted the difference. I could go on at length about similar things, but these will be enough for now.

---

[249] Translator's note. Hieron I of Syracuse, son of Deinomenes, the brother of Gelon and ruler of Syracuse in Sicily from 478 to 467 BC.

Se altro mi chiederete lo faro volentieri le misure di capi[250] come sono colonne tonde, quadre et aguze di piu faccie sperice et simili capi et tenute di vasi et simile sono materie piu aspre a trattar pur quando a voi

16r

dilectassi potro ricorvele[251] dubito no[n] poterle dir se no[n] come dixono gli antiqui et loro le dixono in modo ch[e] co[n] faticha et cognitione di mate maticha et a pena si comprehendono dicovi ch[e] molte cose lasciovi et no[n] dissi bench[é] fussino dilectevoli solo p[er]ch[é] no[n] vedevo modo di poterle dir chiar et ap[er]te come cercavo dirle et in queste durai faticha et no[n] pocha ad exprimerle et farvile intender[e].[252]

---

[250] Translator's note. Cfr. Grayson: 'Le misure de' **corpi**,...'.

[251] Transcriber's note. Difficult to interpret; Translator's note. Cfr. Grayson: '...ricorvele...'.

[252] Translator's note. Cfr. Grayson: '...ad esprimerle e **farmi** intendere'.

If you want to ask more of me, I would be happy to take it further. The measures of items, such as square, round or pointed columns, with several faces, spherical[253] and the like, are topics more difficult to deal with.

16r

If it be to your pleasure, however, I could go over them. I doubt that I can say anything about them any differently than as the ancients did, and they said it in such a way that even with effort and a knowledge of mathematics they are barely understandable. I say to you that many things I have left out and not said, even though they are quite delightful, only because I couldn't see any way to say them clearly and openly as I have sought to say them, and in this I worked quite hard to express them and make them understandable to you.

---

[253] Translator's note. Luca Pacioli used the term *spero* for sphere in *Tractatus geometrie. Summa de Arithmetica et Geometria, Proportioni et Proportionalita* (1494). Leonardo da Vinci also uses the term in his *Libro di Pittura*, for example: '*V. Della parte più oscura de l'ombra ne' corpi sperici o columnali...*' [Bk. V, §752].

Fol. 1r of the manuscript Galileaiana 10 of the Biblioteca Nazionale Centrale in Florence.
Reproduced by kind permission.

Notes on the Manuscript and Criteria for the Transcription of Galileana 10

Angela Pintore

The codex catalogued as Galileiana 10 (Gal. 10) of Biblioteca Nazionale Centrale in Florence, is a paper manuscript dating to the sixteenth century, composed of 23 sheets measuring approximately 21 x 30 cm, today bound and numbered in pencil in the upper right margin. Fol. 1r carries the name of the manuscript's owner, Ostilio Ricci (Fermo, 1540 – Florence, 1603), a court mathematician in the Grand Duchy of Tuscany, best known for having introduced young Galileo Galilei to the study of mathematics and geometry.

The text of Leon Battista Alberti's treatise is contained in fols. 1r-16r of Gal. 10, while the sheets that follow, written in a different hand, are notes of different kinds: fols. 17r–18v contain an inventory dated 1631, while fols. 19r–23r contain mathematical notions and calculations. The transcript of Alberti's brief treatise is complete and its text and illustrations are comparable to the other manuscripts known. However, the lack of the dedication of the treatise to Meliaduse d'Este, as well as any indication at all regarding authorship, led to its being only lately identified as an Albertian codex: Gal. 10 was catalogued in the late 1950s by Angiolo Procissi with the title 'Ricci Ostilio. Problemi geometrici',[1] but was recognised as a transcription of *Ludi Matematici* only some years later by Thomas B. Settle,[2] who from this discovery was also able to draw important implications for the history of science.

In the transcription presented in these pages, we have chosen to remain as faithful as possible to the text of the manuscript, without adding punctuation or accents (which are almost totally lacking in the handwriting of the copyist). For the same reason, the abbreviations have been completed by means of the insertion of square brackets containing the missing letters, in order to render the text more easily readable while at the same time making evident all attempts at interpretation.

In cases where interpretation of the text was doubtful by the insertion of corrections or by the use of one or more individual characters in a word that were not to be found elsewhere in the manuscript, the difficulties in the transcription have been footnoted. Similarly, footnotes have been inserted to note the probable correct interpretation of obvious errors on the part of the copyist, as well as words which appear clearly enough written, but whose meaning is unclear and casts a doubt on the correctness of the interpretation, even where it is not possible to attribute this to errors of transcription.

---

[1] [Procissi 1959].
[2] [Settle 1972].

Notes on the Translation of
Ex ludis rerum mathematicarum

Kim Williams

In spite of the many shortcomings attributed to the 1973 Grayson edition of *Ludi Matematici*, it remains the text of reference at present. (Indeed, given the lack of any kind of punctuation in the Galileiana 10 manuscript we used as the basis of this present translation, including the conventional use of capital letters to indicate the beginning of sentences, the work of interpreting the manuscript in order to translate it would have been infinitely more tedious without the Grayson edition.) Therefore, footnotes have been inserted into the Gal. 10 transcription to indicate places in the text that differ from the edition of Grayson. Such differences may be omissions, additions, or differences in word order. Differences in spelling ('anchora' for 'ancora', 'altezza' for 'alteza', or 'et' for 'e') , where the word remains the substantially the same, have not been footnoted; such differences remain to be noted in a critical edition.

Where Grayson's additions were deemed helpful to an understanding of the text, these have been integrated into the translation in bold characters and enclosed in square brackets [ ]. My own additions to the text for purposes of clarity are inclosed in angle brackets < >.

Because footnotes to the manuscript have been inserted by both the transcriber, Angela Pintore, and myself as translator, these are distinguished as 'Transcriber's notes' and 'Translator's notes'.

Angela Pintore scrupulously maintained the exact lines and page layouts in her transcription, but unfortunately the page layout requirements of this present volume made it impossible to replicate that here. We have, however, clearly denoted folio numbers.

The longer I studied the differences between the Grayson version of *Ludi* with the transcript of the Galileiana 10 manuscript, the more I became convinced that what we are seeing is not the work of a copyist as much as it is of a stenographer; in other words, I think that the manuscript wasn't copied, but rather dictated. What are considered to be errors of copying (by transcriber Angela Pintore, but also by scholars such as Francesco Furlan) appear to me more likely to be the result of dictation, or rather diction. This would explain differences in spelling of words that are quite common ('anchora' for 'ancora', and 'el dardo' for 'il dardo'). It would also explain the insertion of words such as 'cioè' ('that is') as explanative; and differences in phrases that mean the same thing (such as 'as you wish' for 'whatever').

Reading a paper by Francesco Furlan convinced me even more of this idea. He says, for instance, that one of the manuscripts has strong inflections of the area in the Po Valley, which are missing in any other manuscript [2001: 156]; other instances ('dugento' for 'duecento'; 'mena' for 'come') are typical of Tuscany. To me this can be explained by verbal transmission; that is, the reader came from the Po Valley. It would also explain why in the key problem XVII (Bologna-rose-cherry) the reference to a specific place (Monestiriolo) has been changed to simply 'monastery'; that is, the reader, from a different region, doesn't know that this refers to a specific place named after a monastery, and perhaps thinks it is just a way of saying monastery in a local dialect, so he says monastery. Furlan also says that this particular work of Alberti's was – inexplicably – not subject to Alberti's usual careful control of copies of his works [2001: 148]; if the work was transmitted by dictation (imagine a delightful Renaissance evening reading *Ludi* aloud by the fire while a stenographer copies it down), Alberti wouldn't have been there to control it. This may explain why Gal. 10 does not contain the introduction dedicatory paragraph; it may further explain why the illustrations differ from manuscript to manuscript (they would have been drawn on the basis of the text at a different moment). Finally, it explains gaps and lapses: those would be where the stenographer either lost his place or didn't understand.

In any case, these reflections led the conclusion that we should not intervene (even minimally except with notes) in the transcription of Gal. 10, as such interventions in effect constitute a 'version' rather than a 'transcription'. Instead, my translation into English is clearly a version (in the sense that all translations are interpretations). We decided that the reader would be best served if the translation were as literal as possible to allow for comparison between the original and the translation, though this sometimes results in English sentences where the word order is a little awkward.

Commentary on *Ex ludis rerum mathematicarum*

Stephen R. Wassell

Introductory remarks

Despite its Latin title, *Ex ludis rerum mathematicarum* was written by Alberti in the vulgar, and is most often referred to it by its Italian title, *Ludi matematici*, or simply *Ludi*. Alberti wrote most of his other treatises on art or mathematics in Latin, e.g., *De re aedificatoria, De pictura, De statua, De componendis cifris, Descriptio urbis Romae, Elementa picturae*.[1] It therefore stands to reason that *Ludi* is a different kind of text. Indeed, one could interpret it not so much a treatise as a personal favour, written for Meliaduse d'Este at his request, providing a wide variety of applications of mathematics to the real world.[2] It is interesting that Alberti chose the Latin word *ludus* (Italian, *ludo*, plural *ludi*) for the title, a term that 'oscillates between game and spectacle, but also includes without any stretching of the term the concept of exercise, with or without competitive intentions'.[3] Alberti calls one example in the *Ludi* a delightful toy, and describes others as amusing or pleasurable. At times he seemingly tries to impress Meliaduse. At all times he tries to make the text as accessible as possible, presumably because of the fact that Meliaduse was not expert in mathematics.

Ludi was late in being delivered to Meliaduse, by Alberti's own admission in the dedication (a translation of which is included below). Luigi Vagnetti calculates that *Ludi* was over a decade in the making, counting from Alberti's promise to Meliaduse in 1438

---

[1] Two of these treatises were also written by Alberti in Italian, namely *Della pittura* and *Elementi di pittura*. *De lunularum quadratura* seems to be the only other art or math treatise other than *Ludi matematici* written solely in Italian.

[2] Meliaduse (ca. 1400–1452) was one of Niccolò d'Este's many illegitimate children, and was born to the same mother of his younger but more influential brother, Leonello d'Este, who was the Marquis of Ferrara and patron to Alberti. Meliaduse was directed into ecclesiastical life against his wishes by his father, becoming the Abbot of Pomposa. Although Meliaduse was interested in applications of mathematics, he was considered less than an expert and rather a dilettante in this regard. Cf. [Mancini 1882: 195] and [Vagnetti 1972: 175]. Regarding Meliaduse's request of Alberti and a description of the *Ludi*, see [Mancini 1882: 317–318].

[3] Cf. [Vagnetti 1972: 177]: *oscilla tra giuoco e spettacolo, ma include anche e senza forzature il concetto di esercizio, con o senza intenzioni agonistiche* (trans. Kim Williams). Vagnetti's native Italian commentary of the *Ludi* is quite informative, and while we do not ascribe to every one of his views, we feel that one contribution we are making is providing an English forum to discuss some of Vagnetti's insights.

and its completion sometime between 1450 and 1452, and he offers several potential reasons for Alberti's tardiness. Vagnetti questions whether Alberti's heart was completely in the project, given the fact that Meliaduse's lack of mathematical training restricted the content that Alberti was able to include. Perhaps Alberti delayed the project in hopes that Meliaduse would eventually forget about it, or perhaps he was aware of Meliaduse's 'inconsistent dilettantism' and finally satisfied his request only in deference to his social position.[4]

It is quite possible that Alberti's preoccupation with earning and maintaining the patronage of Leonello d'Este was what kept Meliaduse from his satisfaction for so long. As Marquis of Ferrara, Leonello was the most powerful member of the Este family from 1441 until his death in 1450. Anthony Grafton makes a compelling argument that Alberti consciously set out to court the patronage of Leonello through a very measured series of steps. This courtship coincides with the time period leading up to Leonello's assumption of the leadership of Ferrara upon his father's death, and Alberti's successful strategy landed him a position in the court of the new marquis. Meliaduse, Leonello's older brother, first met Alberti presumably in 1435 while Meliaduse was in Florence. Grafton argues that Alberti had identified Meliaduse as one of four men who would eventually help him attract Leonello's patronage.[5] So perhaps Alberti politely delayed writing *Ludi* from 1438 for about a decade, during which time Alberti benefitted from Meliaduse's friendship to establish himself as a prominent figure in Leonello's court, until Alberti felt it was about time to make good on his promise.

In any case it is quite plausible that Alberti viewed *Ludi* as a project meant for a different audience than his normal readership, perhaps only a private audience consisting of the Este family. Indeed, though it is by no means a proof, the fact that it is heavily illustrated suggests that Alberti originally conceived of the *Ludi* as destined to stay out of the public eye. Let us delve into the logic behind this statement.

In the Introduction to the present text, we discuss what Mario Carpo has called 'Alberti's almost obsessive concern regarding errors in the manuscript tradition of his works', and how this was a major reason for the relative absence of illustrations in his written works.[6] But we have in the *Ludi* some two dozen illustrations or more that we can be fairly convinced are based on drawings made originally by Alberti.[7] The conclusion, as

---

[4] See [Vagnetti 1972: 253–255] for more details on potential reasons for Alberti's delay.

[5] See also [Grafton 2000: 207ff]; Grafton mistakenly refers to Meliaduse as Leonello's younger brother.

[6] [Alberti 2007: 17]; Carpo and Furlan argue these issues convincingly and eloquently in their two halves of the introduction to their critical edition and translation (by Peter Hicks) of *Descriptio urbis Romae*. See both our Introduction and Robert Tavernor's Foreword to this present volume for expanded discussions.

[7] See also [Alberti 2007: 23]; it is informative to list some pertinent points written by Furlan about the many illustrations of the *Ludi*: that it is 'a unique case in the ensemble of Alberti's works'; that while the illustrations are by no means uniform in the manuscripts, not even in their number, 'it is nevertheless true that all the manuscripts have figures, that they usually have more than twenty of them,...and that the figures themselves are not always totally simple and elementary'; that 'in the majority of cases the text refers specifically to a figure, although the indications and referrals can

Francesco Furlan states, is that the *Ludi* 'is more a private, or at least a semi-private, document rather than a work destined for dissemination'.[8] In any case, whether or not Alberti intended *Ludi* to be reproduced, it did garner considerable attention, certainly enough to give rise to many manuscript copies, as well as eventual publication in 1568, less than a century after his death, one benefit of the quickly growing world of printing.

While there are many illustrations, and while they do certainly help, it must be stated that some of the games can still be confusing. In some cases the figures themselves can be deceiving. In other cases, while the figures are clearly correct, more mathematical analysis is needed than what Alberti provides in the written descriptions. Further, there is the problem that in the manuscript tradition both the illustrations and the written word are susceptible to corruption in the process of copying and/or transcribing dictation. Of course, Meliaduse himself would have presumably received the original, uncorrupted version from Alberti directly but, corrupted or not, the treatise requires sound quantitative reasoning skills in order to understand all of its mathematical content. Indeed, some of the mathematics can be quite rich, some quite subtle, as we shall see below.

It is also important to note that, in one way or another, all of the measurement techniques Alberti offers are approximations. For one thing, whenever humans take measurements, except when counting objects, there is inherent approximation.[9] More to the point, while most of the games are exact mathematically, at least in theory, some are themselves approximations. Moreover, even for the exercises that are exact mathematically, the methods employed, such as lining up points (in three-dimensional space) by sight, are inherently prone to error. It seems there is an implicit assumption that Meliaduse should expect to get only approximate results. Presumably, though, this would be all that is necessary in many applications.

Regardless of the exact or approximate nature of each game, the most important thing to comprehend is what Alberti is evidently trying to accomplish. Broadly speaking, his intention is to demonstrate to Meliaduse some amazing and entertaining things one can do with mathematics. Consider Alberti's introduction to what is now numbered as Problem 17, certainly the most discussed game (as explained in the commentary below):

> And I want to give a way that with three cherries you can measure how far it is
> in a straight line from Bologna to Ferrara.

This shows a sort of 'amaze your friends' attitude that Alberti conveys in the *Ludi*. This notion in in keeping with the informal tone that Alberti has chosen, from the title itself to the fact that it was written only in the vulgar, to the actual Italian wording and phrasing he uses.

As will be seen in this commentary, the two primary sources for *Ludi* appear to be *De architectura* by Vitruvius and *De Practica Geometrie* by Fibonacci, the latter of which Alberti credits by name (i.e., Leonardo of Pisa). In addition, we must always assume that

---

differ from one manuscript to another'; and thus that 'it is in a manner of speaking probable that the iconographical or illustrative elements really do, in this case, derive from Alberti'.

[8] [Alberti 2007: 23].

[9] To use more precise mathematical language, discrete quantities can be measured exactly, while continuous quantities – such as heights, widths, lengths, depths, and weights – must be rounded to the nearest chosen unit, according to whatever accuracy is required or desired.

Euclid's *Elements* is an important source for Alberti, given the existence of his heavily annotated personal copy.[10] Most of Alberti's potential uses of Euclid involve proportions, which often arise from the well-known fact that corresponding sides of similar triangles are proportional (*Elements* VI, Prop. 4).[11] Vagnetti has speculated on other potential sources never credited by Alberti, such as a little treatise entitled *De Visu*, by Grazia de'Castellani, or an anonymous *Trattato di Geometria Practica*, most likely by a fifteenth-century Florentine, or the works of Marriano Taccola.[12] Grafton also relates *Ludi* not only to Taccola but also to humbler 'abacus books' studied by Italian schoolboys, and he makes the point that 'Alberti mastered many standard instruments and their applications' and 'carried out observations of the sun's path with the Florentine medical man and astronomer Paolo Toscanelli'.[13] Alberti himself names only two other sources besides Fibonacci, Savasorda and Columella, and these will be discussed in the commentary to Problem 12 on measuring fields.

As for those influenced by Alberti, the list is vast and still growing today. Our primary interest in Alberti stems from the combination of his architectural work together with the significant role that mathematics plays in his oeuvre.[14] The same could be said of Andrea Palladio, another prominent Italian Renaissance architect who is often analyzed together with Alberti in this regard.[15] As such, we must admit our interest in the influence of Alberti on Silvio Belli, a sixteenth-century mathematician and engineer called by his colleague Palladio 'the most excellent geometer in our area' – in the general area of Vicenza and presumably more specifically in the membership of the *Accademia Olimpica*, Palladio and Belli being among the founders of the academy.[16] In 1565 Belli published *Libro del misurare con la vista* (*Book on measuring by sight*), which, as one might expect, shares many topics with Alberti's *Ludi*.[17] Belli explicitly cites the work of Alberti in the context

---

[10] Alberti's annotated copy of Euclid's *Elements* is the subject of a recent study [Massalin and Mitrović 2008], which includes as an appendix transcriptions of all of Alberti's annotations and reproductions of his drawings.

[11] See [Euclid 1956: v. 2, 200ff]; further references to this proposition, of which there are several below, will not be cited. Another elementary fact implicit in the problems involving similar triangles is that two angles of a triangle determine the third, which is part of *Elements* I, Prop. 32 [Euclid 1956: v. 1, 316ff]; we choose not even to reference this fact below, except in Problem 16, where Alberti makes a particularly clever use of it.

[12] See [Vagnetti 1972: 183]; Vagnetti also mentions the now well-known thirteenth-century notebook by Villard de Honnecourt, but states that he does not think Alberti was aware of it.

[13] [Grafton 2000: 83–84].

[14] We would like to acknowledge the important role in bringing the authors of this book together played by the biennial *Nexus* conference on the relationship between architecture and mathematics, first held in 1996, and the related *Nexus Network Journal*.

[15] We cite two of the more prominent of these, [Wittkower 1998] and [March 1998].

[16] Palladio's remark is contained in a 1570 letter; Palladio and Belli had worked together on at least one project, namely the Basilica in Vicenza.

[17] Belli first describes the construction of a *quadrato geometrico*, an instrument comprising a square grid with moveable arms for recording points, lines, and therefore angles. Most of his measuring techniques are then described both with and without the use of the *quadrato geometrico*. Of course, the techniques that do not use this instrument are of more interest within the context of this book, since they are more in line with Alberti's methods for Meliaduse. In particular, the interested reader

of the so-called *bolide albertiana*, as we discuss in the commentary to Problem 8. A few other, more prominent names will come up in the commentary, including Leonardo da Vinci in Problems 13 and 19, but it is beyond the scope of this book to examine the impact Alberti had in any comprehensive way.

No.	Description	Location in Gal. 10
1	Measuring the height of a tower, knowing the distance to its foot as well as its height up to a certain point	fols. 1r-1v
2	Measuring the height of a tower, knowing only the distance to its foot	fols. 1v-2r
3	Measuring the height of a tower, knowing only the distance to its foot, using a mirror or a pan of water	fols. 2v-3r
4	Measuring the width of a river from one bank	fols. 3r-3v
5	Measuring the width of a river from one bank that is flat for a distance at least as wide as the river	fols. 3v-4r
6	Measuring the height of a tower, only the top of which is in view	fols. 4r-4v
7	Measuring the depth of a well to the water line	fols. 4v-5r
8	Measuring the depth of deep water	fols. 5r-6r
9	Constructing a fountain powered by gravity, i.e., Heron's Fountain	fols. 6r-7r
10	Determining noon using the sun	fol. 7r
11	Determining time at night using the stars	fols. 7r-7v
12	Measuring the areas of fields of various shapes	fols. 7v-9v
13	Constructing an *equilibra* for leveling surfaces and weighing objects	fols. 10r-11r
14	Measuring the weight of heavy loads	fols. 11r-11v
15	Using an *equilibra* to aim a bombard	fols. 11v-12r
16	Constructing a circular instrument for measuring angles in order to draw the map of a city and to estimate distances	fols. 12r-14r
17	Measuring the distance to a remote but visible place	fols. 14r-14v
18	Constructing a device for measuring lengths of distances along a road	fols. 14v-15r
19	Constructing an instrument for measuring the speed of a ship using the wind	fols. 15r-15v
20	Measuring volumes of solids by displacement of water à la Archimedes	fol. 15v

Table 1. List of problems

will be able to find in [Belli 1565] methods virtually identical to Problems 2, 4, 5, 7, 8, and the last part of 16. Another work by Belli that may be of interest is *Della Proportione, et Proportionalità, Communi Passioni del Quanto* (*On Ratio and Proportion, The Common Properties of Quantity*) [Belli 2003], which deals more with pure mathematics and relates heavily to Euclid's *Elements*.

Let us now consider the games Alberti presents. While he did not title or number them, it has become standard to speak of twenty problems. It is not quite as standard, however, as to which specific exercises correspond to which numbers. Our numbering is shown in Table 1, where we have chosen titles in our own words to describe the content of each problem.

Our numbering is very close to Vagnetti's, identical from Problem 6 onwards.[18] Pierre Souffrin has a somewhat different numbering scheme. Specifically, what we call Problems 4 and 5 become one, our Problems 10 and 11 become one, and our Problem 13 is split into three by Souffrin:

- The *equilibra*, a practical instrument used to measure all kinds of things;
- Levelling and regulating the course of water;
- Weighing minor loads.[19]

Note that all of these numbering schemes agree from Problem 14 onwards.

A few remarks are in order on some of the figures we provide in this commentary. Many of the illustrations in the available manuscripts of the *Ludi* are grossly out of scale, and the manuscript we used and here transcribe for the first time, Galelieana 10 of the Biblioteca Nazionale Centrale di Firenze, is no exception. For example, the arrows or spears shown are often nearly as tall as towers or as wide as rivers. Such distortion is usually not an issue, since the illustrations can be viewed as schematics, which are useful even if not drawn to scale. In the case of Problem 17, however, it has led to incorrect conclusions about the operability of Alberti's method; see the commentary to this problem below. In any case, where we have provided our own illustrations for Problems 1, 2, 3, 6 and 17, we have drawn them to scale, according to Alberti's chosen numbers, as much as possible, in order for the modern reader to get an idea of what it would be like to put these exercises into practice. Finally, where we include and refer to the Grayson illustrations, we mean those that Cecil Grayson includes in his widely circulated edition of the *Ludi*; these illustrations are from the Riccardiani 2942 manuscript.

Dedication

Since the Gal. 10 manuscript does not contain the dedication to Meliaduse appearing in many others, it does not appear in our translation. For completeness, however, an English translation based on the the Grayson edition is provided here:

> I know that with this booklet I am late in satisfying your wishes. And although
> I could give many excuses and reasons for this lateness of mine, I trust to your
> compassion and understanding and ask you for forgiveness if I have done

---

[18] Vagnetti's numbering is based on paragraphs, and there are two differences between his paragraphs and our problems. His third paragraph consists of what he considers to be two special cases of his second paragraph, whereas we feel that only the first of these is really a special case, which we include with our Problem 2. Our Problem 3 is therefore only the second part of Vagnetti's third paragraph. As for the second difference in numbering schemes, what he calls the fourth paragraph is considered by us to be the end of Problem 3 (where Alberti looks ahead to Problems 6 and 17, or possibly 17 and 18). Vagnetti's fifth paragraph is then a combination of our Problems 4 and 5. After that, our problem numbering coincides with his paragraph numbering.

[19] Cf. [Alberti 2002: 26]; Souffrin does not explicitly number the problems, but his list consists of twenty items.

wrong. Perhaps I will have satisfied you, when in these most amusing things collected here you have taken delight, both in meditating on them and also in putting them into practice and using them. I took pains to write them in a very accessible way; and yet I should point out to you that these are very subtle matters, and unfortunately it can happen that they are dealt with so dully that it is hardly worth the effort to pay attention to see what they contain. If they please you, I shall be very happy. And, should you desire anything else, just let me know and I will make every effort to satisfy you. For now I hope that you are pleased with this, in which you will find very rare things. I am entrusting to you Carlo, my brother, a man most dedicated to you and to your family. Fare well.[20]

We have already mentioned Meliaduse's relationship to Alberti, where we also discussed various potential reasons for Alberti's tardiness in satisfying the older brother of his patron, Leonello. Here the most important aspect of the dedication is Alberti's implicit indication that Meliaduse is not expert in mathematics.

Mentioned at the end of the dedication to Meliaduse is Carlo, who, like Leon Battista, was an illegitimate son sired by Lorenzo Alberti, a Florentine merchant. Carlo was 'also a gifted scholar and writer' – both brothers were well educated by their father – but he was not as driven to scholarly pursuits as Leon Battista was.[21]

Problem 1
Measuring the height of a tower, knowing the distance to its foot as well as its height up to a certain point (Gal. 10 fols. 1r-1v)

If we label the eye point as D (fig. 1), as Alberti does in Problem 2, and we call a, b, and c the points on the tower analogous to A, B, and C on the spear, then a straightforward use of two pairs of similar triangles yields the desired result, namely

$$\frac{AB}{BC} = \frac{ab}{bc}.$$

Fig. 1. Our schematic for Problem 1[22]

---

[20] For the Italian text, see [Alberti 1973: 133].
[21] [Grafton 2000: 33].
[22] This and all figures shown here are by Kim Williams unless otherwise noted.

For example, two pairs of similar triangles can be chosen so that the result is obtainable via Euclid's *ex aequali* (*Elements* V, Def. 17 and Prop. 22), a technique that Alberti would have known, given his first-hand knowledge of the *Elements*.[23] To be explicit, since triangles *ABD* and *abD* are similar, it follows (*Elements* VI, Prop. 4) that $AB : BD :: ab : bD$, and since triangles *CBD* and *cbD* are similar, $BD : BC :: bD : bc$. The desired result, $AB : BC :: ab : bc$, is then obtained by applying *Elements* V, Prop. 22 to these two proportions.[24]

In Alberti's example *ab* is 100 feet, *cb* is 10 feet, and he concludes that *BC* will go into *AB* 10 times, nine times into *AC* and the tenth *BC* itself. Note that Alberti is treating the height of the tower as a known, when in practice it would be the unknown sought. Alberti often takes this approach with his numerical examples in the *Ludi*.

To create a schematic that is reasonably to scale and uses Alberti's tower height, we choose the eye point to be 5 feet off the ground and 200 feet away from the tower. This allows us to keep the spear relatively realistically sized: point *A* is 10 feet off the ground, so that the spear needs to be at least as long. This is still quite a long spear, though not nearly as long as the one shown in Gal. 10, which is almost as tall as the tower itself.

Towards the end of the exercise, Alberti cautions that one must keep one's eye at a fixed point, and he advises using a plumb line with (or in place of) the spear. Already we see a general fact about the *Ludi*, namely that while the mathematical methods are sound in theory, the accuracy in practice relies on the skill of execution. Clearly such methods may be used, in any case, to get decent approximations for general needs, as we have discussed above.

Problem 2
Measuring the height of a tower, knowing only the distance to its foot (Gal. 10 fols. 1v-2r)

Although this game may seem more difficult than the first, since now there is no known reference height on the tower, the mathematics is conceptually easier, as it involves only one pair of similar triangles. Again *b* and *c* are the points on the tower analogous to *B* and *C* on the spear.

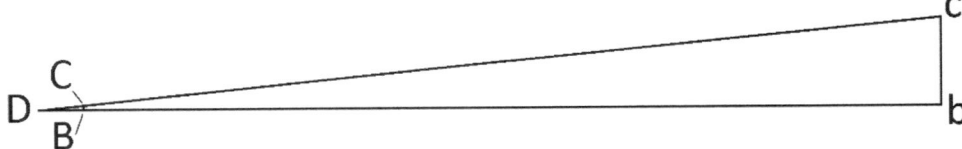

Fig. 2. Our schematic for Problem 2

---

[23] See [Massalin and Mitrović 2007: 199–200] for Alberti's multiple annotations of Euclid's *Elements* Book V, for example.
[24] See [Euclid 1956: v. 2, 179ff]; in modern notation the fact used here is that, given *x*, *y*, *z* and *u*, *v*, *w*, with $x : y :: u : v$ and $y : z :: v : w$, then $x : z :: u : w$.

Triangle CBD is similar to triangle cbD, and therefore

$$\frac{BD}{CB} = \frac{bD}{cb}$$

(*Elements* VI, Prop. 4). In his example Alberti gives cb as 100 feet and bD as 1000 feet, so that CB will go into BD 10 times. Alberti then states that this can be done for any numbers and 'without any error'; rather than cautioning Meliaduse about accuracy, this time he is stressing the correctness of the mathematics, at least in theory. To draw our schematic (fig. 2), we take BD and CB equal to 50 feet and 5 feet, respectively. Alberti's values were probably chosen to make computation easy, not to arrive at a convenient schematic.

This method for finding heights is equivalent to one described in chapter 7 of Fibonacci's treatise, *De Practica Geometrie*;[25] since Alberti mentions 'Leonardo Pisano' by name in Problem 12, in which he uses methods described in chapter 3, we can safely assume that Alberti was aware of chapter 7 as well. Fibonacci gives other methods, as well, including the use of a quadrant (or oroscope), which Alberti does not include.[26] A quadrant is a device for measuring angles, and is essentially half of a modern protractor (while a protractor spans 180°, a quadrant spans only 90°, i.e., one quadrant of the *xy*-plane).

Fol. 1r of part III of *Liber tertius de ingeneis ac edifitiis non usitatis*, Taccola shows an illustration of this method – or a special case thereof, described below (fig. 3). Moreover, fol. 32 demonstrates the same method not for finding the height of a tower but for surveying a mountain (fig. 4).[27] In both illustrations what appears to be a quadrant is lined up along the hypotenuse of the large triangle. In fact, in both illustrations the positioning of the quadrant matches the fact that the triangle shown is an isosceles right triangle, which is the special case that Alberti refers to at the end of this game.

In order to make both triangles DCB and Dcb in fig. 2 right isosceles, which is to say 45°-45°-90° triangles, Alberti instructs Meliaduse to lie flat on the ground with his feet at the bottom of the spear and to sight to a point on the spear as tall as Meliaduse's eye level. Using this approach, one need not even do any mathematics, since the length of line segment bD would be precisely the same as the height of the tower (in theory, of course). An analogous method to produce a right isosceles triangle is explained in Fibonacci's treatise, as well.[28] The right isosceles triangle is an important object in geometry, as Alberti knew, and this special case shows his knowledge of an application thereof.

---

[25] See [Hughes 2008: 346–347 (chapter 7, paragraphs 1 and 2)].
[26] See [Hughes 2008: 349ff (chapter 7, paragraph 5ff)]; Fibonacci describes both the construction and the use of the quadrant.
[27] See also [Prager and Scaglia 1972: 70, 132]; Taccola's *De ingeneis* is dated 1433.
[28] See [Hughes 2008: 347 (chapter 7, paragraph 3)].

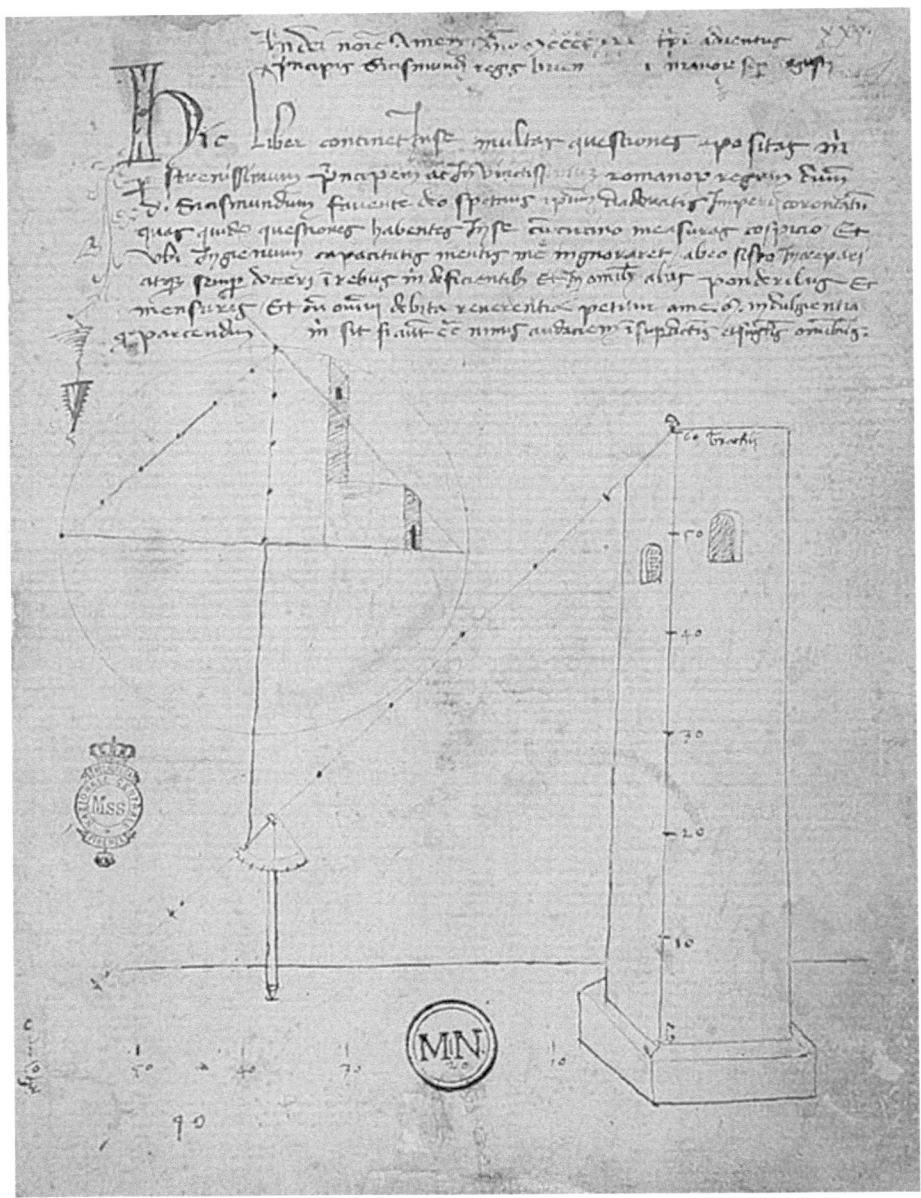

Fig. 3. Taccola, *Liber tertius de ingeneis ac edifitiis non usitatis*, fol. 1r from ms. Palatina 766, Biblioteca Nazionale Centrale di Firenze. Reproduced by permission

Fig. 4. Taccola, *Liber tertius de ingeneis ac edifitiis non usitatis*, fol. 32 from ms. Palatina 766, Biblioteca Nazionale Centrale di Firenze. Reproduced by permission

Vagnetti mentions Villard de Honnecourt's now well-known treatise but surmises that Alberti was not aware of it.[29] We show an illustration from Villard's treatise for the sake of completeness and the interest of the modern reader (fig. 5). This figure seems to be the special case, both from the close approximation of the triangle to right isosceles and the fact that the 45°-45°-90° triangle is quite prominent in the illustration. Villard's triangle is freehand and thus is not exactly right isosceles, so it may be simply the general case of Problem 2, but it looks much more likely that the special case was intended.

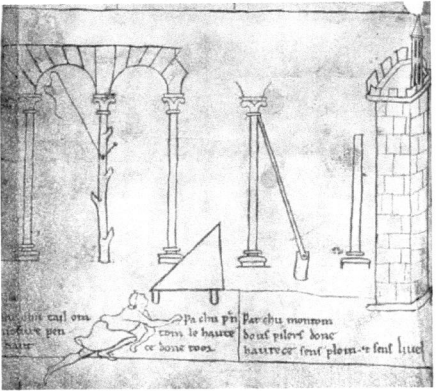

Fig. 5. The Album of Villard de Honnecourt, detail of fol. 20v

Since the approach of Problem 2 to measuring a tower's height is more widely known than the one in Problem 1, especially in the special case, as well as being more direct mathematically, it is natural to wonder why Alberti placed it second. One reason may be because it is less convenient, in that the practitioner must locate his eye as close to the ground as possible. (Note that in the drawing of this method by Villard de Honnecourt, the viewer is shown crouching in a conveniently positioned depression in the terrain.) Another reason is that Alberti presents the tower height problems in order of decreasing knowledge: the tower in Problem 1 is accessible and has a feature the height of which can be determined; in Problems 2 and 3, the tower is accessible but has no such measurable feature; in Problem 6, the tower is not even accessible but only visible.

Unfortunately, Alberti does not address the role of the terrain, which might be the best reason to put Problem 1 first, at least from the standpoint of general practicality. In Problem 2 as well as 3, the terrain between the viewer and the tower is important. Ideally it would be a completely flat horizontal plain (as it is always illustrated, not surprisingly). In fact, this ideal situation is necessary in the special case, in order to make the 90° angle. In the general case, what is crucial is that a true straight-line distance for line segment *bD* must be obtainable (with point *B* also being on this straight line). Problem 1, on the other hand, is operable regardless of the terrain; one simply needs to have a clear view of the whole tower. Thus the key piece of information in Problem 1 – the given height somewhere on the tower – does make for a more widely operable method.

---

[29] See our note 12 and [Vagnetti 1972: 183].

Problem 3
Measuring the height of a tower, knowing only the distance to its foot, using a mirror or a pan of water (Gal. 10 fols. 2v-3r)

Next Alberti inserts into the game a reflecting device, either a mirror or a pan of water, which is to be placed on the ground between the viewer and the tower. This creates two similar triangles in a reflective orientation, namely triangles *ABC* and *DEC* shown in fig. 6 (the illustration from Gal. 10 is not very convincing, as the eye is too high, but these triangles are meant to be similar). The height of the tower can then be found by solving the proportion (*Elements* VI, Prop. 4)

$$\frac{AB}{CB} = \frac{DE}{CE}$$

for the one unknown *AB*.

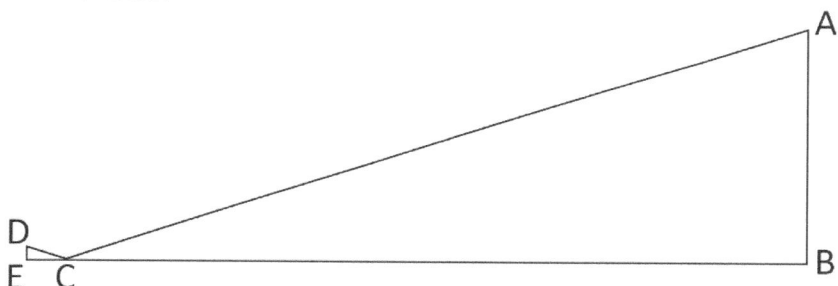

Fig. 6. Our schematic for Problem 3

As with Problems 1 and 2, Alberti's numerical example assumes that the height of the tower, here *AB*, is known. We adjust one number in his example in order to produce a reasonable schematic: we keep *AB* at 100 feet but change *CB* from 1000 to 300 feet. We then take *DE* to be 5 feet (eye level), and so *CE* is 15 feet.

Alberti does not explicitly mention the key fact that the angle of incidence, *ACB*, equals the angle of reflection, *DCE*, but he was well aware of this having been proven. As he states in *De pictura*, the Latin version of *On Painting*, 'as mathematicians prove, reflection of rays always takes place at equal angles'.[30] Some mathematical details and philosophical discussions included in the Latin version of *On Painting* are not included in the Italian version, *Della pittura*. In the *Ludi*, written by Alberti for Meliaduse only in Italian, it is not surprising to see such details omitted.

One could use all of the first three methods for a single tower, to get some helpful checks and balances, though Alberti never indicates this. Instead, Alberti ends Problem 3 with a mention of later games, on 'measuring any distance, especially when it isn't very far away', as well as on measuring 'those quite far way' for which he 'will give [Meliaduse] a singular way'. Presumably the games he foreshadows are Problems 6 and 17, though it is possible they could be Problems 17 and 18. We will quote in full, and comment on, Alberti's foreshadowing in our conclusion to this commentary.

---

[30] [Alberti 1991: 46 (I, xi)].

Problem 4
Measuring the width of a river from one bank (Gal. 10 fols. 3r-3v)

Alberti interrupts the discussion of measuring tower heights for two exercises addressing river widths, each of which uses two spears. In the first river width game, Problem 4, the smaller of two similar triangles is suspended between the spears, and the larger triangle crosses the river (fig. 7). Points C and F are at eye level, and point H is found by sighting from point F to some landmark on the other side of the river, point G. As such, triangle HBG is similar to triangle HCF, albeit at different orientations, so that the desired river width BG can be the unknown in the proportion

$$\frac{BG}{HB} = \frac{CF}{HC}$$

(*Elements* VI, Prop. 4). This is correctly described by Alberti, at least in the abstract, before he gives the numerical example.

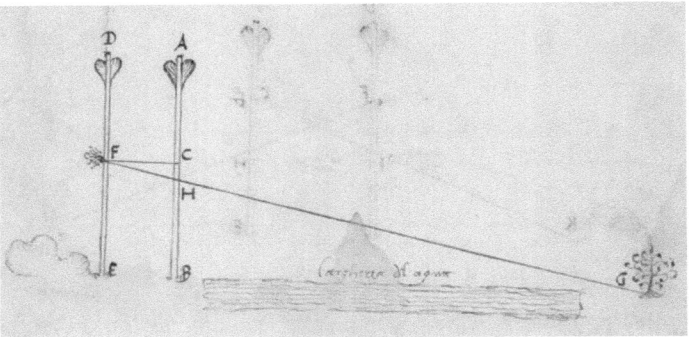

Fig. 7. Gal. 10 illustration for Problem 4

There seems to be some confusion in the numerical example, however. Alberti once again supplies the quantity sought, in this case the river width being 30 paces. The problem seems to be in the specification that FE and CB are each 1 pace, whereas the rest of the numerical example works better (whether or not the insertion from Grayson is used) if instead HB is 1 pace. In any case, the game should be clear to those familiar with similar triangles, despite the potential confusion.

There is a third triangle similar to the aforementioned two, namely triangle FEG, at the same orientation as triangle HBG. If we interpret letting 'the river be thirty paces wide' as EG being 30, then the rest of the numerical specifications would work: FE equals 1 pace, as specified, and then HC would indeed go into CF thirty times. This seems to be a bit of a stretch, though.

The usual assumption is necessary here, namely that points E, B, and G all lie on the same line. Alberti actually explicitly states at the beginning of this game that Meliaduse should '[p]lace [himself] with [his] feet in a flat place' and it is clear that this requirement would need to extend to the other side of the river as well. Then since line segments FE and CB are equal (both being the height of the practitioner's eye), line segment CF will be parallel to line segment BG, which is all that is required to produce similar triangles.

Problem 5
Measuring the width of a river from one bank that is flat for a distance at least as wide as the river (Gal. 10 fols. 3v-4r)

If the river bank where the spears are located is flat for a distance at least as wide as the river (and at the same elevation as the opposite side), then one can reproduce the width of the river directly on the bank without solving any proportions. To explain the method, Alberti expands upon the construction given in Problem 4. (Note that in fig. 8, what was correctly labelled point E in fig. 7 has now mistakenly been dropped.) He first instructs Meliaduse to locate point I on the second spear at a distance from F the same distance as H is from C on the first spear. Meliaduse then must carefully place his eye at point C, as shown, and aim through point I towards the ground of the river bank. Where the visual ray hits the ground, point K is marked, so that the distances BK and EG are equal (or perhaps more to the point, the distances EK and BG are equal).

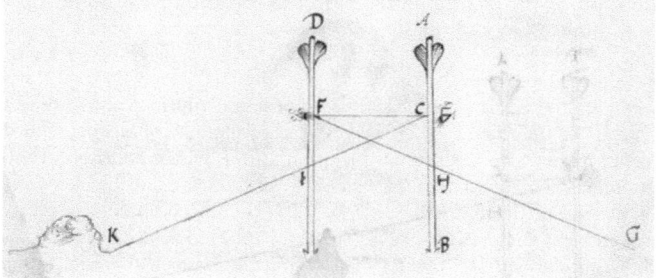

Fig. 8. Gal. 10 illustration for Problem 5

Villard illustrates an instrument that allows for a similar method as that given in Problem 5, which Vagnetti calls a *grafometro di confronto* (fig. 9).[31] The two arms pivot independently, and the practitioner aims, from a given position on this side of the river, both arms by sight to a point on the other side. Rotating 180° at the same given position, he can reproduce the distance to the point on the other side of the river, taking advantage of the clean slate afforded by the flat bank. We do not have any evidence that Alberti knew of Villard's notebooks, but they were written two centuries before.

Fig. 9. The Album of Villard de Honnecourt, detail of fol. 20r

---

[31] [Vagnetti 1972: 200].

Problem 6
Measuring the height of a tower, only the top of which is in view (Gal. 10 fols. 4r-4v)

Alberti may have placed this game after Problems 4 and 5, and not with Problems 1 through 3 concerning tower heights, because it requires more sophisticated mathematical thinking. Indeed, he prefaces Problem 6 by stating that it is 'marvellous' yet 'laborious to understand'. The level of mathematics is high enough that it may have confused some transcribers along the way; below we shall note some definite problems in Gal. 10. In any case, the mathematical content of the exercise is recoverable beyond any reasonable doubt.[32]

The game in Problem 6 is to measure the height of a tower by remotely sighting its top from two separate points, marking the intersection of the sight line on the spear from each point. In the figure the top of the tower is labelled C, the base of the spear B, the two eye points D and G, and the two respective intersections on the spear E and F (fig. 10). Alberti's explanation of the mathematics is not as detailed as it could be, perhaps because he did not want to overwhelm Meliaduse.

---

[32] We first wish to note that while [Furlan 2006] discusses corruptions concerning Problems 6 and 17 in various manuscripts of the *Ludi*, it does not provide mathematical analyses for these two games in specific detail, as compared for example with [Furlan 2001], which addresses Problem 17 rigorously from both philological and mathematical bases. We avail ourselves of the excellent analysis provided by [Furlan 2001], as well as [Furlan and Souffrin 2001], in our commentary to Problem 17 below. In any case, the mathematical analysis we provide here for Problem 6 is completely our own.

While [Furlan 2006] concerns more directly the relationship between text and figure in the 'technico-scientific' works of Alberti, within the context of the supremacy of the written word over illustrations from antiquity through the Middle Ages, it is instructive that Furlan argues the potential supremacy of illustrations when it is known that they derive from the author; see [Furlan 2006: 206]. We feel it appropriate, then, to revisit Furlan's own conclusion with regard to the *Ludi*: 'it is in a manner of speaking probable that the iconographical or illustrative elements really do, in this case, derive from Alberti' [Alberti 2007: 23] (see note 7 in the introductory remarks to this commentary for quotations regarding the logic Furlan uses to arrive at this conclusion).

Let us return to the game at hand. While Problem 17 certainly benefits from the rigorous scholarly apparatus Furlan has constructed (see our commentary to that problem for details), we maintain that Problem 6 does not require a strict philological approach to understand the mathematics fully. The major difference between Problems 6 and 17 is that, while in Problem 17 not only is the text clearly corrupted, so are the illustrations – and in different ways with different manuscripts, with Problem 6 the illustrations agree and are almost certainly correct, given Alberti's overall description of the problem. Since the mathematics can clearly be reconstructed well beyond any reasonable doubt, a rigorous philological approach, though always beneficial, does not seem to be necessary, at least if one is primarily interested in understanding the mathematical content of the game.

Regarding the text, in Problem 6 (and with the *Ludi* in general) our approach in this book is to offer both a transcription and an English translation of a manuscript that has not before been available in published works, i.e., Gal. 10, with clearly marked comparisons to, and selected insertions from, the widely available (albeit philologically flawed) Grayson edition. We then provide in this commentary our best analysis of the mathematics, commenting both on the historical aspects and the modern understanding, as appropriate. We look forward to seeing a critical edition of the entire *Ludi*, but we do not feel it necessary to wait for such a publication to offer our own careful and rigorous mathematical analyses.

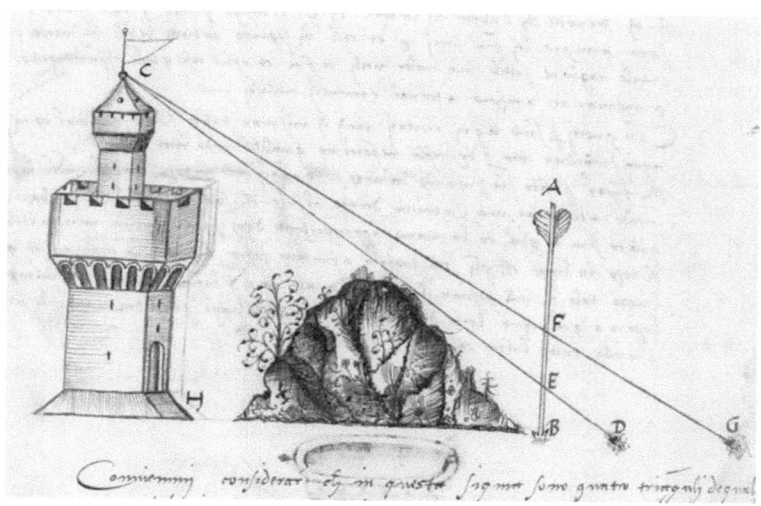

Fig. 10. Gal. 10 illustration for Problem 6

He starts the mathematical explanation by pointing out that there are two pairs of similar triangles, *FBG* similar to *CHG* and *EBD* similar to *CHD*. The next line of the text, however, probably has suffered from corruption, since it seems to compare *GH* from the first pair of similar triangles with *DB* from the second.[33] Even more problematic is the numerical example that follows. Alberti's use of the proportions $EB:DB::CH:DH$ and $FB:BG::CH:HG$ to obtain the ratio $HG:DH$ is correct in theory, but in the numerical example $EB:DB$ is taken to be 2 and $FB:BG$ is taken to be 3, which is impossible: it is clear from the construction that $EB:DB$ must be greater than $FB:BG$. This could simply be a matter of mixing up the order of division, e.g., whether *DB* goes into *EB* or vice versa. This mix-up could have been caused by a corruption, especially considering the fact that it is then undone by another mix-up, namely that the ratio $HG:DH$ is found to be $3:2$, when in fact it would be $2:3$ given the erroneous choices of $EB:DB$ and $FB:BG$. In any case, now that the two mix-ups have cancelled out each other, the explanation is essentially back on track, with the finding that the ratio $HG:DH$ is $3:2$. Of course, as Alberti then points out, we do not know either quantity since the tower is inaccessible.

Here is where the more sophisticated mathematics comes into play. Although Alberti does not reference Euclid (he never does in the *Ludi*), the procedure corresponds to 'conversion' of ratios (*Elements* V, Def. 16 and the corollary to Prop. 19).[34] In modern notation the fact is that $u:v::x:y$ implies $u:(u-v)::x:(x-y)$; here we assume $u>v$ and $x>y$. Correspondingly, Alberti's numerical example, $HG:DH::3:2$, implies $HG:(HG-DH)::3:(3-2)$, i.e., $HG:DG::3:1$, the last proportion being precisely

---

[33] As noted in the *Ludi* translation, this problem persists whether or not one chooses to incorporate the Grayson text differences (see note 36, p. 21 in this present volume).

[34] See [Euclid 1956: v. 2, 115 and 174ff].

what Alberti concludes. At this point, Alberti takes his usual approach of pretending the tower height is known (and, in this case, conveniently equal to DG) in order to provide a readily understandable example for Meliaduse. Alberti then avoids getting into more details, and it seems unlikely that the methods of this exercise would be useful to Meliaduse (or anyone else) without putting in a substantial amount of additional analysis.

It would have been more helpful for Meliaduse's use if Alberti had explained more completely the last few steps of the process in the practical case where the tower height is unknown, although to be fair, in Alberti's time there probably was not sufficient notation developed to do so succinctly. Now we can easily let $x = EB/DB$ and let $y = FB/BG$. Then $HG/DG = x/(x-y)$, which allows one to find HG (since DG is directly measurable); HG can then be used in the proportion $FB : BG :: CH : HG$ to find CH, the height of the tower. A compact formula arising from this modern analysis would then be

$$CH = \frac{xy}{x-y} DG \,.$$

It is easy to take for granted how useful good notation can be! In any case, we want to stress that the analysis contained in the current paragraph is ours and is not indicated in the least by Alberti. We provide it simply for the benefit of the modern reader.

For our schematic we maintain Alberti's ratio $CH : HG$ as 1:3 (with the former 100 feet and the latter 300), but we cannot maintain the equality between CH and DG, since this would force a ridiculously tall spear (fig. 11). We take $DG = 12$, $DB = 12$, $EB = 4\ 1/6$, and $FB = 8$, so that the spear would need to be at least 8 feet tall.

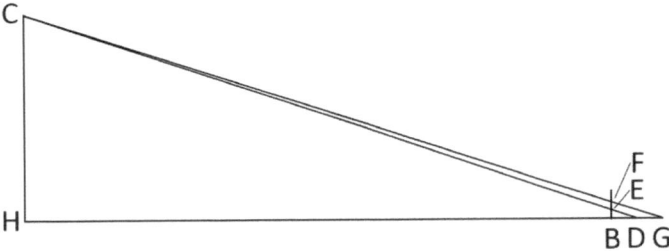

Fig. 11. Our schematic for Problem 6

In Souffrin's French translation and commentary of the *Ludi*, Problem 6 is the first game on which he comments.[35] He discusses 'false position' (*fausse position*) techniques, wherein one solves a system of linear equations by making an initial (but incorrect) guess at the unknown quantity and then adjusting as needed to find the correct solution. He cites a 1343 work by Jean de Murs that incorporates this technique in an example, and he casts the numbers of Alberti's game into an analogous example. While his approach is completely different than ours, both underscore the fact that Problem 6 has a rather large hole at the end.

---

[35] See [Alberti 2002: 83–87] for the details, which are certainly interesting from a mathematical and historical point of view.

Problem 7
Measuring the depth of a well to the water line (Gal. 10 fols 4v-5r)

The illustration from Gal. 10 for Problem 7 is potentially confusing, in that it shows only three horizontal levels. Other illustrations of Problem 7, as in the one from the Grayson edition (fig. 12), clearly show four different horizontal levels: the very top of the well (corresponding to point C), the level of the reed (line segment AB), the water line (corresponding to point D), and the bottom of the well (which has no designation). The fact that the last of these is not shown in the Gal. 10 illustration causes confusion in that point D could be mistaken for the very bottom of the well (figs. 12, 13).

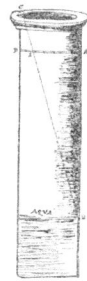
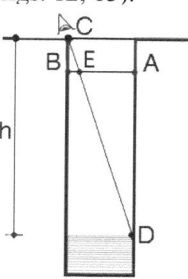

Fig. 12. Grayson illustration for Problem 7   Fig. 13. Illustration of Problem 7
(from *Ludi* manuscript Riccardiana 2942,1)   after [Vagnetti 1972: 205]

In any case, the reed wedged in the well along a diameter is the key to obtaining similar triangles, and for the mathematics to work, the reed needs to be placed horizontally (parallel to the water line). The visual ray from point C to point D intersects the reed at point E, which produces similar triangles BEC and AED. Alberti's conclusion, however, is not a direct application of the usual fact for similar triangles (*Elements* VI, Prop. 4). Only a minor additional step is needed – really two steps, if we restrict ourselves solely to the repertoire available in Euclid's *Elements*, as we will now show.

A direct application of the aforementioned similar triangles yields the proportion

$$\frac{AD}{BC} = \frac{EA}{EB}$$

(*Elements* VI, Prop. 4). One could then use the 'composition' of ratios (*Elements* V, Def. 14 and Prop. 18).[36] In modern notation the fact is that $u : v :: x : y$ implies $(u + v) : v :: (x + y) : y$; applying this to the proportion above yields

$$\frac{h}{BC} = \frac{AB}{EB}$$

(note that we use the fact that $AD + BC = h$, the depth of the well, in the numerator on the left side; on the right side, we use $EA + EB = AB$, the length of the reed). We now simply need to take the 'alternate' ratio (*Elements* V, Def. 12 and Prop. 16) of both sides to obtain

---

[36] See [Euclid 1956: v. 2, 115 and 169ff]; Heath also uses 'componendo' for 'composition'. This corresponds to 'conjunction' in the Hughes translation of Fibonacci; see, e.g., [Hughes 2008: 57].

$$\frac{h}{AB} = \frac{BC}{EB}$$

in order to match Alberti's words completely.[37] This analysis is fairly minor – certainly not as complex as Problem 6 or even Problem 1 – and Alberti suppresses it.

The numerical example starts in the usual way, with Alberti providing the quantity sought (he takes the depth of the well to be 21 *braccia*), as well as a second quantity (the length of the reed, 3 *braccia*). He does not provide a third (let alone a fourth) quantity, but rather leaves the example hanging a bit at the end:

> Thus you will find, measuring as I have described, that EB goes into BC as many times as there are reeds that measure your well. I won't go into measuring these depths here, because I know that you have the wits to understand all things that are similar to this.

Here, 'as many times' would simply be 7, so it is curious that he does not just say this. In any case, Alberti seems to express confidence that Meliaduse would be able to fill in the details, if he so desired.

Problem 8
Measuring the depth of deep water (Gal. 10 fols. 5r-6r)

This game is the first non-geometric problem, the first that uses what we shall call an engineering approach.[38] Alberti suggests the use of a water vessel with a hole in the bottom to measure the passage of time, and he also describes the construction of a specialized floater and sinker. The idea is to force the sinking of an object that otherwise floats – what we are calling a floater – by attaching a sinker that dislodges once it hits the bottom of the body of water, thereby allowing the floater to return to the surface. Meliaduse can time this process with various known depths, in order to develop a scale by which to convert the amount of water dripped out of the vessel to the depth of the body of water.

Alberti's floater is an oak gall, i.e., a ball-shaped growth from the bark of an oak tree, which is inherently bouyant. The sinker is a plumb, i.e., a lead weight usually used to weight a string in order to produce a perfectly vertical line of reference (as he suggests using in various games contained in the *Ludi*). As with Problem 7, the Gal. 10 figure in Problem 8 is problematic (fig. 14); here the deficiency is the depiction of the gall, which, if we are reading the figure correctly, is shown as a very small ball at the top of the lambda-shaped plumb. In the Grayson illustration for Problem 8 (fig. 15), the gall is shown separately from the plumb, so that the 'minute wire made similar to an abacus figure that stands for 5' can be shown attached to the gall as instructed. The illustration after Vagnetti (fig. 16) shows how the shape of the plumb allows point C to touch first, then point B, allowing the gall to dislodge and float back to the surface.

---

[37] See [Euclid 1956: v. 2, 114 and 164ff].

[38] See [Grafton 2000: 77ff] for an excellent discussion on what could be called the advent of *bona fide* engineering, in the context of works by Brunelleschi, Taccola, Leonardo da Vinci, etc., and which offers etymological considerations of the term 'engineer' (*ingeniator* or *engignour* being related to the concept of *ingenium*, meaning ingenuity as well as an ingenious device).

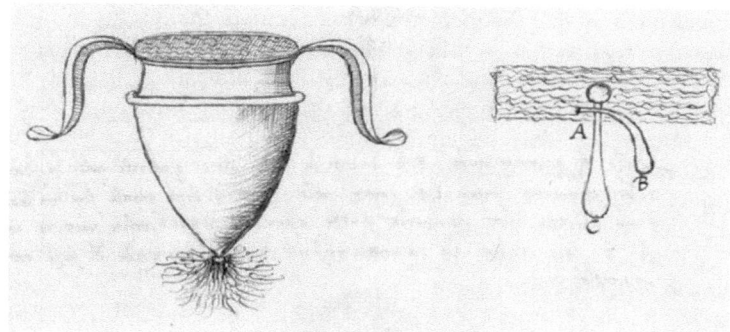

Fig. 14. Gal. 10 illustration for Problem 8

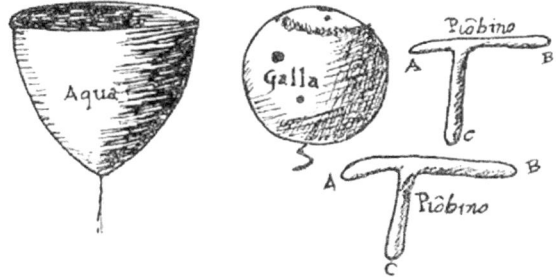

Fig. 15. Grayson illustration for Problem 8 (from *Ludi* manuscript Riccardiana 2942,1)

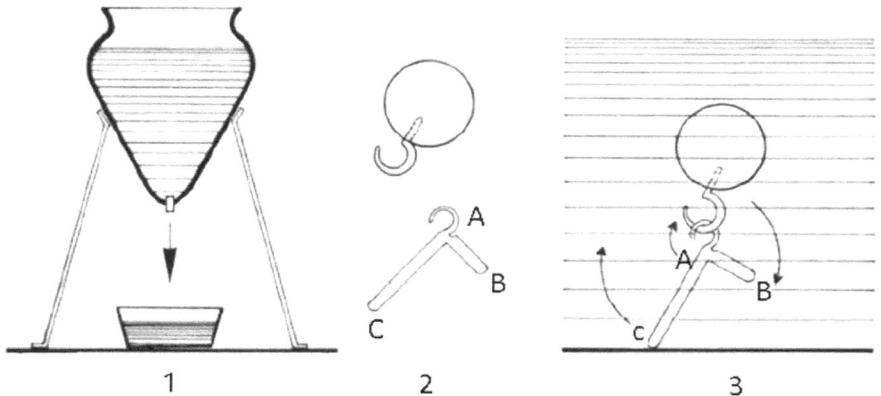

Fig. 16. Illustration of Problem 8 after [Vagnetti 1972: 208]

Alberti gives a caveat that the method works as long as there is no current in the water, and so he seems to realize that its practicality is somewhat suspect. Perhaps he desires to justify the usefulness of the game anyway, by then indicating that the water-dripping vessel can function as a clock, regardless of the context. Vitruvius wrote about water clocks at length in the final chapter of Book IX.[39] Alberti then states that anything exhibiting motion (by which he clearly means motion with constant velocity) can be used to measure time. Souffrin makes the point that any cultivated man of Alberti's time would recognize in this statement a fundamental theory of Aristotle: *time is the name of movement*.[40] Alberti also mentions the possibility of making clocks with fire (oil candles), as well as air and water (the next game).

Earlier in his transition from Problem 7 to Problem 8, Alberti refers to the latter as 'a certain way described by the ancient writers'. It is natural to wonder which ancient writers he is referencing. Mancini instructs that Alberti was most likely referring to Savasorda, an eleventh- and twelfth-century Jewish mathematician and astronomer born in Barcelona, who is explicitly described as an ancient writer by Alberti in Problem 12 (see our commentary below).[41] Mancini notes that, in place of an oak gall, Savasorda uses a very light ball of copper or lead, and that Savasorda also suggests two methods of timing the process, using either the water dripping method indicated by Alberti or an astrolabe.

Mancini also points out that several authors after Alberti referred to the device as a *bolide albertiana*, and some of these authors comment on potential inaccuracies of the method due to differences in water density depending on the salinity of the water, as well as the variation in the rate of water dripping out of the vessel depending on the volume of water in the vessel. Vagnetti repeats much of this analysis, but he also cites works by other Renaissance authors who were aware of Savasorda's description.[42]

Vagnetti notes the relationship of the *Ludi* to Silvio Belli regarding this game.[43] Belli includes the floater and sinker method in his 1565 book *Libro del misurare con la vista* (*Book on measuring by sight*). Most of Belli's methods involve similar triangles, such as we have seen in previous problems, but he closes his book with what he describes as two beautiful and creative propositions from books of most excellent men, about the first of which he writes: 'This about measuring the depth of the sea I have read in a book written in pen on measuring of Leon Battista Alberti of Florence'.[44] Belli's rehearsal of Alberti's method is complete with the gall, the wire in the shape of the number 5, the lambda-shaped plumb, and the water vessel for timing purposes.

---

[39] See [Vitruvius 2009: 265ff (IX.xiii)].

[40] See [Alberti 2002: 92], which also contains an extended discussion about average velocity versus instantaneous velocity, in the context of the Latin *velocitas* as used in medieval and Renaissance times.

[41] See [Mancini 1882: 321–322].

[42] Cf. [Vagnetti 1972: 207–209].

[43] See [Vagnetti 1972: 210]; also, see the introductory remarks to this commentary for a discussion of Belli, p. 78 and note 17 in this present volume.

[44] Cf. [Belli 1565: 104]: *Questa del misurare la profondità del mare l'ho letta in un libro scritto a penna del misurare di Leon Battista Alberti Firentino* (trans. Kim Williams). Note that 'written in pen' indicates that Belli had consulted a manuscript rather than a printed book.

Problem 9
Constructing a fountain powered by gravity, i.e., Heron's Fountain
(Gal. 10 fols. 6r-7r)

Recall that the meaning of *ludo* includes the concept of show or spectacle, and Problem 9 is a good example of this. Although Alberti does not cite his source, he describes a version of Heron's Fountain, named after its inventor, the Greek first-century CE mathematician and engineer also know as Hero of Alexandria.[45] The fountain is powered by water pressure due to gravity, and today we would call it an example of hydraulics and pneumatics.

Three chambers are placed at different heights; initially the upper and middle chambers are filled with water and the lower one is empty, i.e., filled with air. To start the fountain flowing, a connecting pipe from the upper chamber to the lower chamber is opened, allowing the water to flow. The air displaced by this water proceeds from the lower chamber to the middle chamber via a second pipe, which forces the water out of the middle chamber through a third pipe, producing the fountain.

Alberti describes the construction of a vessel, drawn as a cylinder, comprised of the three chambers stacked on top of each other. The boundary between the upper and middle chambers is called *GH*, and the boundary between the middle and lower is called *EF*. The three pipes are all concealed within the vessel, so that the fountain coming out of the top of pipe *IK* seems to emanate from out of the top chamber, heightening the effect of the spectacle. Pipe *LM* must initially be plugged at *L*, while pipe *NO* is always left open (*N* must be positioned close to the top of the middle chamber). Alberti also notes the need for a hole *P*, which is used to fill the middle chamber before being plugged, at which point the upper chamber can be filled with water. Unplugging *L* then starts the show.

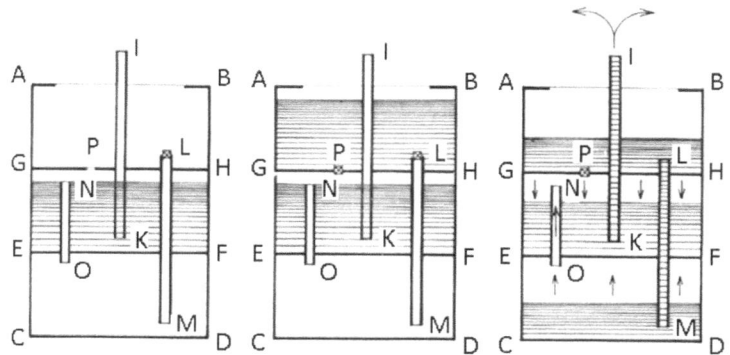

Fig. 17. Illustration of Problem 9 after [Vagnetti 1972: 211]

Alberti skips one detail either in the construction of the fountain or in the description of its use. The missing detail is simply that, once the fountain has run its course, the water in the lower chamber must be drained out before one can repeat the process as described.

---

[45] See [Vagnetti 1972: 212–213].

Alberti's construction is sufficient as long as the fountain can be turned upside down, in order to transfer the water from the lower chamber to the middle chamber (for this purpose, point *O* should be towards the top of the lower chamber, as instructed by Alberti but in contrast to the way it is drawn in the Gal. 10 manuscript). In fact, this would accomplish two tasks at once, simultaneously emptying the lower chamber and refilling the middle one. The other solution, which is necessary if turning the fountain upside down is not feasible, is to add a (pluggable) hole at the bottom of the lower chamber for draining purposes.

Problem 10
Determining noon using the sun (Gal. 10 fol. 7r)

To begin the discussion of the tenth game, Alberti returns to a discussion of clocks. He notes that telling time can be achieved with accurate observation of the sun and stars, using such instruments as the quadrant, the astrolabe, and the armillary sphere. Alberti then makes the point that simply knowing when it is noon is the most basic skill one needs for telling time, and that he wants to present a simple method for doing so using the sun, in the spirit of the *Ludi*. Alberti returns to the use of a spear stuck in the ground, this time in order to cast a shadow from the sun. Vitruvius gives a full account of sundials and water clocks in last chapter of Book IX, but Alberti provides a simplified version.[46] Alberti's statement that his method works in 'any country' is valid, even in the Southern Hemisphere where shadows would be to the north instead of south, as long as the sole purpose is to record where noon is.

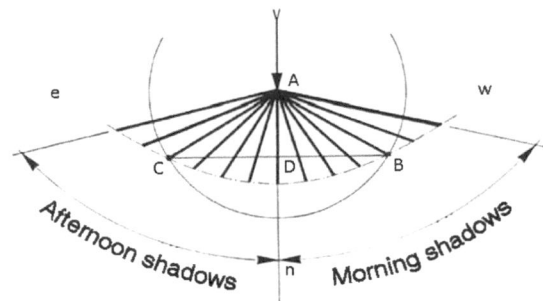

Fig. 18. Illustration of Problem 10 after [Vagnetti 1972: 214]

The method is a nice use of symmetry, as it uses a circle to find the match of a given time before noon with its symmetric time after noon. Alberti instructs Meliaduse to stick a spear straight in the ground in a flat place and record where its shadow falls at some time he knows is before noon. The spear is at point *A* and the tip of the shadow is point *B*, the latter marked with a stick. Then he must take a cord to act as a makeshift compass, using *A* as the centre, and draw a circle on the ground with radius *AB*. As noon approaches, the sun will be higher in the sky and will therefore cast a shorter shadow; after noon, the shadow will lengthen again. When the shadow 'precisely comes to touch [Meliaduse's] circle', he

---

[46] See [Vitruvius 2009: 265–275 (IX, v, 1–14)].

is to label the tip of the shadow as point C; the time will then be exactly as much after noon as it was before noon when the tip of the shadow was at B. This is why Alberti instructs Meliaduse to find the midpoint of line segment BC, which he labels D, and draw the line segment AD, which will point due north (fig. 18).

We feel compelled to register a small complaint concerning Vagnetti's commentary on Problem 10. He points out that the direction of line segment AD could also be determined by a magnetic compass. Of course, this ignores the fact that true north and magnetic north can vary by several degrees – indeed, magnetic north varies over time – but that is not the cause of our complaint.[47] Vagnetti writes at some length about the history of the compass, and given the fact that Alberti does not mention it, neither in this game nor in the next, Vagnetti concludes that the compass was 'evidently not known to him'.[48] This logic is faulty. Just because Alberti chose to illustrate this particular method does not mean that he was unaware of others, or even that he was unaware of the magnetic compass. In fact, it is obvious from *De re aedificatoria*, for example, that Alberti knew about many things that he did not include in *Ludi*. We believe a more reasonable explanation is that Alberti chose this method because it best fit his criteria for mathematical games.

Problem 11
Determining time at night using the stars (Gal. 10 fols. 7r-7v)

This game involves telling time on a clear night by keeping track of how much the constellations have revolved around the North Star, i.e., Polaris (which Alberti calls *tramontana*). One necessary skill is to find Polaris, which Alberti addresses towards the end of the exercise. The other skill is to pick out some other nearby star and keep track of its revolution around Polaris. Reasoning that a whole revolution corresponds to a whole day, Alberti explains that a quarter revolution would correspond to six hours, a third a revolution to eight hours, and so forth. It may require decent visualization skills to avail oneself of this method, but that is really the only limit on the accuracy, besides the weather.

Towards the end Alberti tells of how two stars, what we now call Merak and Dubhe, from the constellation Ursa Major (also known as the Big Dipper or Plough), line up to point the way towards Polaris, which is at the very end of the tail of Ursa Minor (i.e., the handle of the Little Dipper). The line points the direction, and Alberti gives an estimate for the distance from the two stars to Polaris (presumably he means the distance to Polaris from the nearer star of the two, Dubhe). In the Gal. 10 manuscript this distance is estimated to be three times the distance between Merak and Dubhe; instead, in Grayson the estimate is three and a half times. Looking at a modern star chart, we see that both are underestimates; the distance to Polaris is between five and six times the distance between Merak and Dubhe. In any case, Alberti mentions two different names for the end of Ursa

---

[47] According to the U.S. NOAA (National Oceanic and Atmospheric Administration) National Geophysical Data Center website, the magnetic declination in Florence, Italy is almost 2° (1° 59′ E, changing by 0° 6′ E each year). See www.ngdc.noaa.gov/geomagmodels/Declination.jsp. Visited May 12, 2010.

[48] [Vagnetti 1972: 215–216]; see Vagnetti's footnote 78 for his history of the magnetic compass.

Major consisting of Merak and Dubhe: Wheels of the Cart (*ruote del carro*) or Mouth of the Horn (*boccha del corno*).

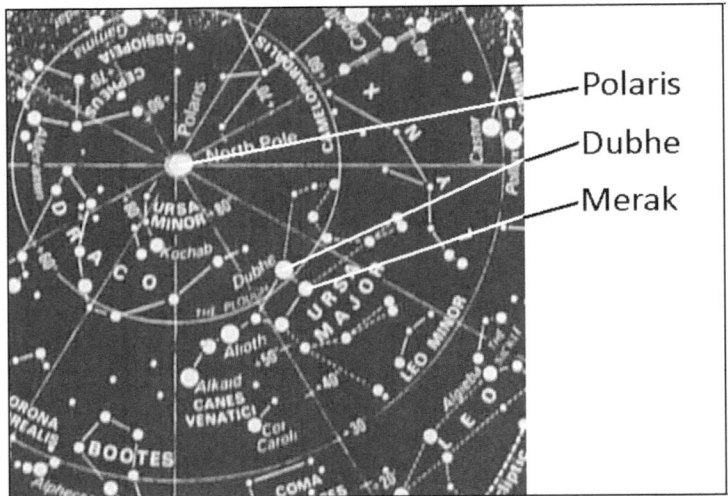

Fig. 19. Star chart showing Ursa Major. Reproduced courtesy of the Astronomical Observatory of Torino

In Alberti's (presumed) autobiography, written in the third person, he writes:

> He wrote little books on painting, and in that art did unheard-of things, incredible to those who saw them, which he showed in a little box with a small opening. ...He called these things 'demonstrations,' and they were such that expert and layman alike would insist that they saw, not painted things, but real ones in nature. There were two kinds of demonstrations, those he called diurnal and those he called nocturnal. In the nocturnal ones you would see the North Star, the Pleiades, Orion, ...[49]

Alberti was quite learned about the night sky, and in fact was a significant astronomical collaborator with Paolo Toscanelli.[50] Another source of knowledge is Vitruvius, where the majority of Book IX is devoted to astronomical and astrological issues.[51] More generally, this game shows how mathematics was understood by Alberti as part of broader field of study. The *quadrivium* (arithmetic, music, geometry, and astronomy) formed the upper division of the seven liberal arts in the medieval university system (the lower division was the *trivium*: grammar, rhetoric, and logic), and the topics of the *quadrivium* were more tied together then than now:

> While nowadays we are used to thinking about music as a purely audible phenomenon, during the Renaissance audible music was regarded only as an aspect of *musica mundana*, the music of the world, encompassing ratios on the

---

[49] [Grafton 2000: 84–85].
[50] See Lionel March's commentary on *On Squaring the Lune*, pp. 209–210 in this present volume.
[51] See [Vitruvius 2009: 249ff (IX, i–vii)].

human body, the movements of the heavenly spheres and *also* what we call music today.[52]

Our modern notion of arts and sciences, and the division of each into disciplines and sub-disciplines, did not begin in earnest until the scientific revolution sparked by the Enlightenment.

Finally, this game is a good example of the problem with illustrations in the manuscript tradition. The stars are available in the sky for all to see, but still the copyist managed to mangle the depiction of Ursa Major.

Problem 12
Measuring the areas of fields of various shapes (Gal. 10 fols. 7v-9v)

'But let us return to what you asked me...' begins Alberti in Problem 12, on land surveying, indicating that Meliaduse had asked Alberti specifically about that topic. It is by far the largest exercise in the *Ludi*, about triple the average length, with more illustrations than any other. It is really a collection of exercises. There can be no doubt that a major source for Problem 12 is *De Practica Geometrie*, Fibonacci's treatise on geometry, in particular Book III, which is also by far the largest book of that collection. Alberti himself cites three people by name at the beginning of Problem 12: '[t]he ancient writers, especially Columella, Savazorda [sic] and other surveyors, and Leonardo Pisano among the moderns'. Leonardo of Pisa is the important twelfth- and thirteenth-century Italian mathematician now known primarily as Fibonacci. Although he is best known in modern culture for the Fibonacci sequence, he was important in the history of mathematics primarily for recovering a good deal of ancient Greek knowledge via the Eastern cultures, where they had been preserved and studied during the Dark Ages of the Western cultures. Fibonacci also transmitted and expanded upon some of the mathematics that had been developed in the East, and in the process he helped introduce the West to Hindu-Arabic numerals, which became standard and are now used by a large percentage of the world's cultures.[53] As far as *Ludi* is concerned, Fibonacci's *De Practica Geometrie* could easily have provided Alberti the basis for much of Problem 12.

Columella was the first-century CE Roman author of *De re rustica*, a treatise on agriculture, and he is 'considered the founder of agronomy'.[54] In the second chapter of Book V of his treatise, Columella offers an example that is almost certainly the source of an example cited by Alberti at the end of Problem 12. The example is a calculation of the

---

[52] [Mitrović 2001: 115]; in this article Mitrović reassesses the 'harmonic proportions' debate sparked by [Wittkower 1998], who had written about the use of proportions in the architecture of Alberti and Palladio being potentially informed by music theory.
[53] See also [Katz 1998: 307–309]. Fibonacci's more famous publication is his *Book of Calculation*, the *Liber abbaci*. Fibonacci begins this treatise by introducing the Hindu-Arabic numerals and the place-value system, and then he gives various examples to show the ease with which one can add, subtract, multiply, and divide whole numbers and common fractions. The *Liber abbaci* is also where the Fibonacci sequence appears, as a solution to a problem about counting the number of pairs of rabbits in a population, given a particular breeding pattern. The sequence is now best known for its connection to the golden section, but this connection was not discussed by Fibonacci.
[54] [Lévy-Leblond 2003: 51].

area of a field which is the shape of a segment of a circle, and we shall consider Columella's formula in some detail below.

Abraham bar Ḥiyya, better known as Savasorda, spelled with a 'z' by Alberti, was an eleventh- and twelfth-century Jewish mathematician and astronomer born in Barcelona.[55] The text of the *Ludi* would suggest that Savasorda, like Columella, was an ancient writer, as opposed to 'moderns' such as Leonardo Pisano (Fibonacci), when in fact Savasorda is closer in age to Fibonacci than Alberti is. Here is a possible explanation. Alberti's source for Savasorda's Hebrew original *Ḥibbur ha-mešiḥah we'tišboret* (*Treatise on Mensuration and Partition*) would almost certainly have been Plato of Tivoli's Latin translation *Liber embadorum* (*Book of Areas*). Moreover, Plato of Tivoli was known for dealing with ancient works like that of the astronomer Ptolemy, which could have lead to some confusion on Alberti's part. As for the mathematics of Savasorda, Alberti is on the mark. Savasorda, a mathematician in the geometrical tradition who draws from Euclid and does not mention the *al-jabr* (the source of our modern term 'algebra') tradition in *Treatise on Mensuration and Partition*, was also one of Fibonacci's sources for *De Practica Geometrie*.[56]

As one would expect, and as is the case in chapter 3 of Fibonacci's treatise, Alberti starts with the most basic shapes of fields, namely squares and rectangles. Of course, Fibonacci goes into much more detail, concerning working with different units of measure, and offering many examples with varying degrees of difficulty – his work is a treatise, as compared to the informal and much shorter *Ludi*. Alberti's sole example of a square is 10 paces by 10 paces, and for a rectangle he chooses 10 by 20. He then proceeds to triangles, but only right triangles, and his example is, not surprisingly, a right isosceles triangle with each leg 10 paces long.

Alberti then suggests that irregular tracts of land should be subdivided into rectangles and right triangles, with the use of a large square (large because it 'errs less'). Alberti then describes a handy construction for a square, namely to form a 3-4-5 right triangle using a string that is 12 units long. This well-known construction is found, for example, in Vitruvius.[57] Its inclusion is in keeping with the theme of the *Ludi*: interesting and useful applications of mathematics. Alberti also takes the opportunity to point out the inexact nature of some people's use of the 5-5-7 triangle, which is not exactly a right triangle. Indeed, checking the Pythagorean Theorem, the sum of the squares of the legs is 50, while the hypotenuse squared is 49. This is why it is correct for Alberti to have written 'one of the fifty parts is missing' (as in Grayson, whereas Gal. 10 has 'five' in place of 'fifty').[58]

Next Alberti discusses circular fields. He uses the standard Archimedean approximation of π, namely 3 1/7 or 22/7, which was to Alberti's time what our standard decimal approximation for π is today, 3.14. Multiplying the diameter by π gives the circumference. This is useful in and of itself, and moreover it is the first step in Alberti's

---

[55] See [Calinger 1996: 53ff]; the name Savasorda is based on a twisted pronunciation of his court title.

[56] See [Hughes 2008, xxiii–xxiv], [Calinger 1996: 54–56] and [Katz 1998: 297–299] for more about Savasorda's influence on Fibonacci.

[57] See [Vitruvius 2009: 244 (IX, introduction, section 6)].

[58] Vagnetti [1972: 223] replicates the error in Gal. 10.

algorithm for finding the area. The second step is to multiply together half of the diameter and half of the circumference, to produce the area. This accords with our modern formula: half of the diameter (the radius, r), times half the circumference (equivalent to r times π), yields $A = \pi r^2$. As usual, Alberti picks an example where numbers work out nicely; he chooses a diameter of 14, which yields a circumference of 44. Half of each of these numbers multiplied together then gives the area, 7 times 22 equals 154. These facts are faithfully recorded in the illustration, nicely arranged within a circle (fig. 20).

Fig. 20. Fourth Gal. 10 illustration for Problem 12

The final shape that Alberti treats is a segment of a circle, the region obtained by slicing off a piece of a circle with a straight line, a shape 'similar to a moon that is half full or less'. Now the story starts to get more complicated; indeed, Souffrin characterizes this last part of Problem 12 as 'the most difficult geometric problem considered in the *Ludi* and, in fact, in all of Alberti's oeuvre'.[59] Alberti first notes the ease with which one considers the half circle, a special case of segment, but then he asks Meliaduse to consider segments less than half a circle, 'similar to a bow'. In fact, a 'bow' (*archo*) becomes Alberti's metaphor for the segment of a circle, and an 'arrow' (*saetta*) becomes his metaphor for the height of the segment (since the arrow of a bow would be positioned at the height of the corresponding segment). It is therefore difficult to know whether Alberti means *corda* to be 'chord' or 'cord', where the latter would be a metaphor for the former. We've maintained 'cord' in the translation to fit the metaphors of 'bow' and 'arrow', but we'll use chord, segment, and height in the commentary. Fig. 22 shows the segment of fig. 21 extended to a full circle, with the height *bd* of the segment extended to an entire diameter *be*.

---

[59] Cf. [Alberti 1998: 91]: *Ce dernier problème est le problème géométrique le plus difficile considéré dans les* Ludi *et, de fait, dans toute l'œuvre d'Alberti* (trans. Stephen R. Wassell).

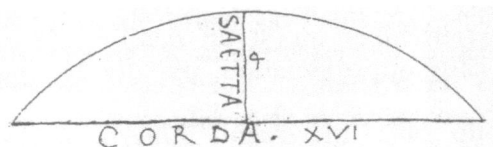

Fig. 21. Illustration for Problem 12 from *Ludi* manuscript Classensis 208 (folio not numbered). Reproduced by permission of the Biblioteca Classensis in Ravenna

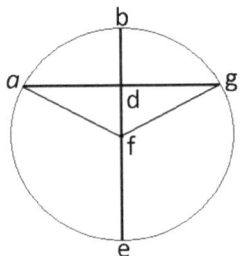

Fig. 22. Illustration for Problem 12 after [Souffrin 1998: 93], which is from Boncompagni's edition of Fibonacci

Alberti's first point on the subject of a segment is that the ancients made a table for determining chords. Indeed, Ptolemy's *Almageste* contains a table of chords, as Alberti would have known from reading *De Practica Geometrie*. There Fibonacci not only mentions Ptolemy's work, he also offers his own table of chords.[60] Continuing, Alberti wavers a bit, seemingly wanting to avoid any explanation by maintaining: 'but these are very intricate things and not suitable for these games I am proposing'. He again suggests that Meliaduse simply subdivide a given field into rectangles and triangles in order to measure its area. As we have seen before, however, Alberti proceeds to go ahead and describe what he hesitates to do at first: 'But if you would still have some principle for understanding their reason…', he states, and then he starts a numerical example to find the area of a segment given its chord and height. Souffrin argues that what appears here in the *Ludi* is at least a partial description of a process to find the area of the segment using a table of chords.[61]

The first step is to find the diameter of the circle. Alberti uses the fact, which he could have known by Euclid's *Elements* III, Prop. 35, that if two chords of a circle intersect (as in fig. 23), then a times b equals c times d.[62] Alberti applies this to the chord and diameter of fig. 22 (the diameter is itself a chord, of course). He takes the chord (4), divides by two (2), squares the result (4), which then equals the product of the height (1) with the remainder of the diameter. This allows him to find the remainder (4), which is then added to the height to get the full diameter (5). After this first step, the diameter, or rather the radius, could be used to find the area of the sector, the pie-shaped piece *fabg* of fig. 22; the

---

[60] See [Hughes 2008: 344–345, 354ff]; Hughes places the table of chords in chapter 7.
[61] See [Souffrin 1998: 94ff].
[62] See [Euclid 1956: v. 2, 71ff].

area of the triangle *fag* could then easily be found and subtracted from the area of the sector to obtain the area of the segment.

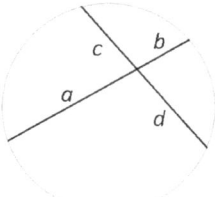

Fig. 23. An illustration of *Elements* III.35, which works for any two intersecting chords

Unfortunately, what is demonstrated in the *Ludi* after the first step is only the computation of the area of the triangle. We recognize the formula that we would now call 1/2 base times altitude (using 'altitude' rather than 'height' so as not to get confused with the height of the segment). Alberti takes the radius of the circle (2 1/2) and subtracts the height (1 in Grayson; 2, incorrectly, in Gal. 10) to get the altitude of the triangle (1 1/2 in Grayson; 1/2, incorrectly, in Gal. 10); this altitude is then multiplied by half the base of the triangle, i.e., the chord divided by 2 (2), resulting in the area of the triangle (3 in Grayson; 2 3/4, incorrectly [and inexplicably], in Gal. 10). What is missing in *Ludi*, as Souffrin points out, is the computation of the area of the sector (from which to subtract the area of the triangle just computed, in order to obtain the area of the segment). The area of the sector is simply the arc of the segment times half the radius, and the arc of the segment is obtainable (at least approximately) from a table of chords, as is found in Fibonacci's *De Practica Geometrie*.[63] Souffrin makes the point that previous commentators on the *Ludi* seem to have been content in assuming that Alberti intended the area of the triangle to serve as an approximation for the area of the segment, which would indeed be quite a rough approximation the vast majority of the time. Of course, the confusion stems from the fact that the numerical example ends prematurely, which Souffrin reasons was caused by corruption in the received manuscripts.[64]

The last exercise in Problem 12 is an alternate way to approximate the area of the segment of a circle, which Alberti attributes to Columella. In the second chapter of Book V of his treatise, Columella offers the following example:

> If [the field] is less than a circle, it is measured as follows. Consider an arc the base of which is 16 feet, and the height 4. The base and the height together make 20 feet. Multiplied by 4 [the height], we obtain 80, the half of which is

---

[63] See [Hughes 2008: 354ff]. Fibonacci's table was based on a circle with diameter 42 rods, so that the user had to employ a three-part process to determine, from a given chord and diameter, the arc of the corresponding segment. The chord is first scaled to the table (multiply by 42 and divide by the diameter); then the table is used to find the nearest chord and read off the arc; and then the arc is scaled back to the original field (multiply by the diameter and divide by 42).

[64] As should be evident by now, we owe much of this analysis to [Souffrin 1998: 94–101]; on p. 100, Souffrin considers the possibility that Alberti either did not know or did not understand how the table of chords could be used to complete the computation, but Souffrin argues that a corruption in the manuscripts, namely a shared lacuna, is more likely.

40. The base being 16 feet, its half is 8. Multiplied by itself, it gives 64, the fourteenth part of which is a little over 4. Added to 40, this gives 44. I say that this is how many square feet there are in the arc [the segment].[65]

This formula can be expressed in modern notation as follows:

$$\frac{1}{2}F(C+F)+\frac{1}{14}\left(\frac{C}{2}\right)^2$$

In contrast, the corresponding formula from Grayson's *Ludi* (incorrectly formulated in [Souffrin 1998: 101] but corrected in [Alberti 2002: 105]), would be expressed as:

$$\frac{1}{2}F(C+F)+\frac{1}{14}\left[\frac{1}{2}F(C+F)+\frac{C}{2}\right]$$

Both Gal. 10 and Grayson thus agree with the 40 found in Columella (even though Gal. 10 is missing a few key words of the calculation). However, the part of the formula in which the *Ludi* is incorrect results in 1/14 times the quantity 40 plus 8; both Gal. 10 and Grayson are adjusted so that Columella's 'a little over 4' becomes 'three and a piece more' instead (which is correct for 48/14). The end result in Grayson is 'about 44' (which agrees with Columella except in that the real answer in Columella is a bit more than 44 while in Grayson it is a bit less). In contrast to Grayson, there is an apparent corruption for the end result in Gal. 10: 54 is obtained, presumably by adding to 40 the number 14 (as opposed to 1/14 of 48). It seems certain that Alberti's intention, at very least, was to cite Columella's example faithfully.[66]

Problem 13
Constructing an *equilibra* for leveling surfaces and weighing objects (Gal. 10 fols 10r-11r)

This game has some interesting mathematical subtleties, as we shall see, but perhaps the most important aspect of this problem for some historians is that it provides relative dating information regarding *Ludi* and Alberti's most famous written work, *De re aedificatoria*. Alberti starts by remembering that Meliaduse had asked him a question about conducting water, and his first response is to refer Meliaduse, for the more complete story of how to control water, to 'those books of mine on architecture' that he wrote at the request of Leonello d'Este, the more prominent brother of Meliaduse. Indeed, in *De re aedificatoria*, Alberti thoroughly discusses the 'four operations concerning water: finding, channeling, selecting, and storing'.[67] Vagnetti points out that the management of water

---

[65] This translation is taken from [Lévy-Leblond 2003: 51]; the Latin can be found in [Souffrin 1998: 101, note 19].
[66] Lévy-Leblond [2003] compares Columella's formula to the true area the segment, and the graph given in his Figure 2 on p. 52 shows that the approximation is close for segments smaller than half a circle. This article also provides an interesting and believable explanation as to how Columella's formula may have been developed in the first place, including the appearance of the surprising number 14, which results from using 22/7 to approximate π.
[67] [Alberti 1988: 325ff (X, iii)].

must have been very important to Meliaduse, since there was a good deal of it on the holdings of the Abbey at Pomposa.[68]

The device constructed for this game, using an arrow, some cord, and a plumb, is what Alberti calls an *equilibra*, a type of balance or scale. Alberti's *equilibra*, which he uses first for the levelling of water and later for the weighing of small objects, is really a makeshift archipendulum. Fibonacci defines an archipendulum in chapter 3 of *De Practica Geometrie*, the same book that is all about measuring fields (see the commentary on Alberti's Problem 12); towards the end of chapter 3, Fibonacci uses the archipendulum as an aid in measuring fields that are not level, e.g., on a hillside.[69] The fact that the archipendulum comes right after measuring fields in the *Ludi* is evidence for the notion that Fibonacci was Alberti's primary source on both of them.

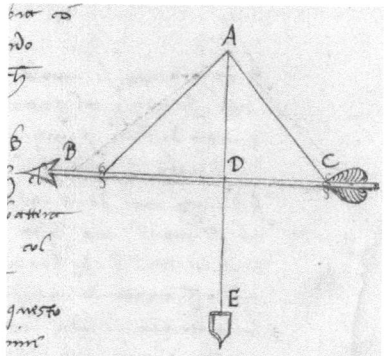

Fig. 24. First Gal. 10 illustration for Problem 13

The construction of the *equilibra* is simple enough, and is fairly evident just from the picture alone (fig. 24). Alberti instructs Meliaduse to take a straight rod, be it a spear or something similar, attach strings of equal lengths to each end, and tie the other ends of the strings together, labelled *A* in the figure. At point *A* is attached a third string with a small weight at its other end, and the whole *equilibra* then hangs from *A*; ideally, of course, the spear would be perfectly horizontal. The point where the weighted string intersects the spear is then marked with a piece of wax, for reference.

As for the uses of the *equilibra*, Alberti first indicates its function as a balance: loads hung from either end can be balanced by trial and error, by ensuring that the vertical string coincides with the wax. He saves an extended discussion of balancing loads for the end of the game, however, shifting instead to the function of the *equilibra* as a level. The application he has in mind is levelling water, which inherently means levelling where the water is contained. This is where the real fun begins, from a mathematical point of view.

Alberti takes the opportunity, in this exercise, to address the fact that the curvature of the Earth can be a factor in levelling water over large distances. Vitruvius, in a quick mention of this fact, cites Archimedes as maintaining 'that water is not horizontal but part

---

of a sphere of which the centre coincides with that of the earth'.[70] Interestingly enough, Alberti uses a different numerical figure for the curvature of the Earth in the *Ludi* (1 foot [or perhaps braccio] per 9000 feet) as compared to *De re aedificatoria* (ten inches per mile).[71] In any case, in this game he describes a clever way to deal with the curvature, which deserves the following extended analysis.

Souffrin describes the method as one that is fitting for the *Ludi* because it allows the user to avoid a fairly significant calculation, and he notes that the method would still be used in the eighteenth century by D'Alembert.[72] While the illustration in Gal. 10 (fig. 25) is acceptable (certainly as compared with Bonucci and Vagnetti, as we shall discuss below) and exhibits the possibility of the eyes being at different levels (an improvement over the corresponding Grayson figure, fig. 26), the fact is that the vertical lines drawn upward from the eyes would actually be diverging, given that they meet in the centre of the Earth.

Fig. 25. Second Gal. 10 illustration for Problem 13

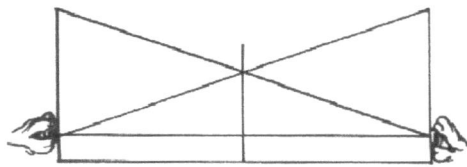

Fig. 26. Grayson illustration for Problem 13 (from *Ludi* manuscript Riccardiana 2942,1)

Souffrin's schematic exaggerates the divergence, as is done in three analogous diagrams from D'Alembert, which Souffrin includes in his commentary.[73] In particular, we should view each of the line segments AM_2 and BM_1 as 'vertical' (i.e., radiating from the centre of the Earth, fig. 27).

---

[70] [Vitruvius 2009: 236 (VIII, v, 3)].
[71] See [Alberti 1988: 336 (X, vii)]. On the ratio from the *Ludi*, [Vagnetti 1972: 229] states that some codices replace the unit of measure *braccio* with the unit of measure foot, and therefore it cannot be established what Alberti wrote.
[72] See [Alberti 2002: 88]; the 1784 D'Alembert work is 'Nivellement' in the *Encyclopédie méthodique – Mathématiques*. While Souffrin's analysis is flawed, as we shall see, there is still a good deal of merit to his commentary on this game.
[73] [Alberti 2002: 91].

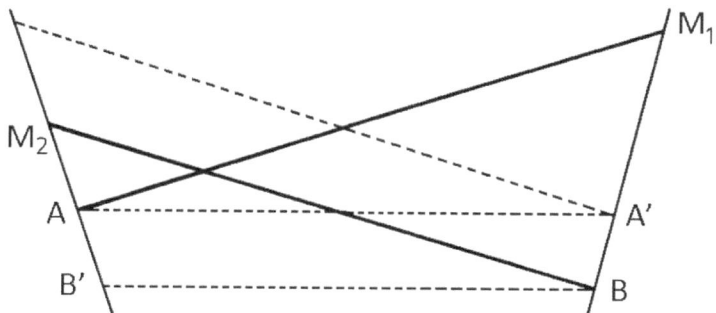

Fig. 27. Schematic for levelling water in Problem 13 after [Alberti 2002: 89]

Let us first relate this schematic to the game at hand. Here is Alberti's first mention of the curvature of the Earth: 'But here many are fooled when levelling, first because they don't understand that the earth is round and turns so that wherever part you are levelling seems to be higher than the other'. It may require a bit of visualization to understand this; we choose to use the second person for the explanation. Suppose you and a colleague are located on one of the world's salt flats, which are sometimes used in attempts to break the land speed record with ultra-fast vehicles, because for vast distances the ground is almost perfectly flat (i.e., at virtually the same elevation, that is, at virtually the same distance from the centre of the Earth).[74] Suppose that you are standing at one end of the salt flats, with your eye located exactly at ground level (pretend this is possible). Your colleague is standing at the other end, as far away as possible on the salt flats, holding a surveyor's pole in a perfectly 'vertical' position. You aim a perfectly 'horizontal' line of sight (or perhaps a laser beam, using a modern surveying instrument such as a theodolite) at your colleague's pole. The fact is that where this 'horizontal' line intersects the pole will be slightly above the surface of the Earth. Why is this? Your 'horizontal' line of sight is perpendicular to the line segment from the centre of the Earth to your eye, which is to say it is tangent to the surface of the Earth. As this line of sight approaches your colleague's pole, it diverges from the surface of the Earth, which curves downward, so to speak. Thus, as Alberti states, 'wherever part you are levelling [i.e., at your colleague's pole] seems to be higher than the other [i.e., your eye on the surface of the Earth]'.

So, how would one obtain the same level (i.e., elevation, or the distance from the Earth's centre) at two points very distant from one another? Alberti instructs Meliaduse to

> level from here to there, and mark the targets [**and then from there to here
> and again mark the targets**] in equal fashion, and of the whole difference take
> the half and this will be suitable measuring for you.

Now we return to Souffrin's schematic (fig. 27). Meliaduse sights from point A on the left 'vertical' (a spear, say), using his *equilibra*, to produce point M_1 on the right 'vertical' spear. Then he sights in reverse, from B to M_2. It is important to note that Souffrin correctly shows A and B at different levels. This is because one cannot assume that A and B

---

[74] The Bonneville Salt Flats in Utah is the usual location in the United States where such activities take place.

are at the same elevation (presumably Meliaduse is not located at the Bonneville Salt Flats, or any other perfectly level place), but the beauty of Alberti's method is that this does not matter, as we shall see. Also note that A' and B' (points at the same elevation as A and B, respectively) are shown for purposes of analysis; they would not be known to Meliaduse, at least not yet.

Now we proceed to Souffrin's analysis, which is mostly correct (and we will address the flaws). First we must interpret Alberti's words: 'of the whole difference take the half and this will be suitable measuring for you'. The natural interpretation (and the correct method, mathematically) is to find the midpoints of line segments AM_2 and BM_1 in order to produce the required points that are at the same level; these are not shown in Souffrin's schematic, but they would be a bit above A and A', respectively. We should point out here that 'the same level' is a bit of an overstatement: Alberti's method is approximate, though it is extremely close, as we shall see.

In order to proceed with his demonstration, Souffrin first labels some pertinent quantities. He sets H equal to the distance from A to B' (equivalent to the distance from B to A'), D_1 equal to the distance from A' to M_1, and D_2 equal to the distance from B' to M_2. Also, he calls the radius of the Earth R (and he implicit assumes that B is on the surface of the Earth, so that R is the distance from the centre of the Earth to B). He then states that, by construction,

$$\frac{D_2 - D_1}{D_1} = \frac{H}{R} \gg 1 \text{ [sic].}$$

There are two problems with this equation. First, the right symbol is backwards: as it is, this would say that H/R is much greater than 1, when clearly H would be tiny relative to R. The second problem is that the subscripts on the left side of the first equation are interchanged. Since we are making this claim, it is best to go ahead and show the steps, which Souffrin suppresses. It is fairly clear, given the various symmetries of the construction, that triangles M_1AA' and M_2BB' are similar.[75] This readily produces the proportion $D_1 : D_2 :: AM_1 : BM_2$. Now consider our modification of Souffrin's schematic, where the lines AM_2 and BM_1 are extended until they meet at the centre of the Earth, C (fig. 28).

---

[75] To prove this, call the intersection of AM_1 and BM_2 point P, and note that triangles M_1BP and M_2AP are similar; indeed, in these triangles, angles A and B are right angles, the angles at P are clearly equal, and therefore the angles at M_1 and M_2 must be equal. It now suffices to show that in triangles M_1AA' and M_2BB', the angles at are A and B are equal. But these are both congruent to the angle at A' between the dashed lines emanating from A', which can be seen by using appropriate parallel lines as well as the overall symmetry of the diagram. As a final side note, realize that while triangles M_1AA' and M_2BB' are similar, they are not congruent, since D_1 will be slightly larger than D_2.

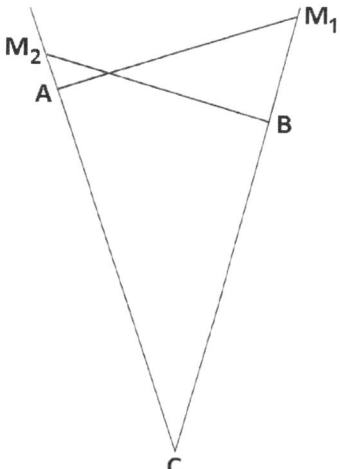

Fig. 28. Our extension of the schematic from fig. 26, showing the centre of the Earth at C

Clearly triangles CAM_1 and CBM_2 are similar (both are right triangles and both share the angle ACB), which leads to the proportion $CA : CB :: AM_1 : BM_2$. Since the right-hand-sides of the two proportions just produced are the same ratio, we can deduce that $D_1 : D_2 :: CA : CB$. But $CB = R$ and $CA = R + H$, so that the last proportion yields

$$\frac{D_1}{D_2} = \frac{R+H}{R}$$

and hence

$$\frac{D_1 - D_2}{D_2} = \frac{H}{R} \ll 1$$

(this last step, subtracting denominators from their respective numerators, is what might be called 'disjunction' of ratios, as opposed to 'conjunction'; see the commentary to Prob. 7 for use of the latter).[76] Now that we have corrected Souffrin's proportion, we can continue with his analysis. Both ratios of the last proportion are positive, and if a positive quantity is much less than 1, it must be very close to 0. Moreover, a fraction is very close to 0 if and only if either its numerator is very close to 0 or its denominator is very large. Since D_2 is not very large, $D_1 - D_2$ must be close to 0, i.e., D_1 and D_2 must be close to each other. From the fact that D_1 and D_2 are essentially equal, Souffrin concludes: 'The distance between the two marks is therefore 2H, which was to be demonstrated'.[77] By 'the two marks' he means M_1 and M_2, but frankly we feel it is best to ignore his conclusion. Instead,

---

[76] We are not providing references to Euclid's *Elements* for this discussion, since we are filling in the details not for Alberti's analysis but for Souffrin's. As for the terms conjunction and disjunction, see [Hughes 2008: 57].

[77] Cf. [Alberti 2002: 90]: *La distance entre les deux marques est donc 2H, ce qu'il fallait démontrer* (trans. Stephen R. Wassell).

recall we are trying to show that the midpoints of line segments AM_2 and BM_1 are at (virtually) the same elevation. Recalling that B and B' are assumed to be at the surface of the Earth, i.e., at elevation 0, the midpoint of AM_2 is at an elevation equal to the average of H and D_2, i.e., $(H + D_2) / 2$, and the midpoint of BM_1 is at an elevation equal to the average of 0 and $H + D_1$, i.e., $(H + D_1) / 2$. Since D_1 and D_2 are essentially equal, so are the elevations of the two midpoints, which we set out to demonstrate.

While Souffrin's method works, at least with the modifications we provide, we shall now present an alternate method that is much more direct and brings into the discussion the relationship between the arithmetic mean and the geometric mean, a topic of interest in its own right; it is addressed by Alberti in *De re aedificatoria*.[78] As stated above, our modified schematic (fig. 28) easily yields a pair of similar triangles, CAM_1 and CBM_2. There is a fairly obvious proportion to form from the similar triangles, in the context of Alberti's method, namely:

$$\frac{CA}{CM_1} = \frac{CB}{CM_2},$$

which leads by cross multiplication to $CA * CM_2 = CB * CM_1$.[79] Taking the square root of both sides (which brings us back to normal length units, as well), leads us to the revealing fact that it is the geometric means that are equal, not the arithmetic means.[80] To understand the import of this, reconsider Alberti's instructions: 'of the whole difference take the half and this will be suitable measuring for you'. As stated above, the natural interpretation is to find the midpoints of line segments AM_2 and BM_1 in order to produce the required points that are at the same level, i.e., points that are equidistant from the centre of the Earth, C. Finding the midpoint of AM_2 is tantamount to taking the average (arithmetic mean) of CA

---

[78] The arithmetic mean of two quantities a and b is simply their average, $(a+b)/2$, while their geometric mean is $\sqrt{ab}$. Alberti writes at some length about the three principal means, the third being the harmonic (or 'musical') mean, in *De re aedificatoria*, within the context of three-dimensional composition. At the beginning of the discussion, Alberti states that the rules for composition of means 'are many and varied, but the wise use three principal methods, whose object is to find, given two other numbers, an intermediate one, which will correspond to the other two by a fixed rule, or, to put it another way, by a family relationship' [Alberti 1988: 308 (IX, vi)]. He then proceeds to give rules for, and numerical examples of, the three principal means, wherein he also notes that the arithmetic is always the largest mean, next the geometric, and then the harmonic. Alberti ends the chapter with the following: 'By using means like these, whether in the whole building or within its parts, architects have achieved many notable results, too lengthy to mention. And they have employed them principally in establishing the vertical dimension' [Alberti 1988: 309 (IX, vi)]. That is, given the length and width of a rectangular room, an architect may apply one of the means to find a suitable height. Palladio includes these same ideas in his treatise, where he also provides illustrations for the constructions of the three means [Palladio 1997: 58–59 (II, xxiii)]. For further discussion about the three principal means, as they relate to classical Greek mathematics, Palladio, and others, see [Wassell 2002: 58–61].

[79] Of course, cross multiplication is covered by Euclid, twice in fact: in *Elements* VI, Prop. 16 for lines (which is geometric, corresponding to magnitudes) and in *Elements* VII, Prop. 19 for numbers (i.e., whole numbers); see [Euclid 1956: v. 2, 221ff and 318ff, respectively].

[80] To be complete, the arithmetic means are equal in the trivial case, when points A and B are at the same level to begin with.

and CM_2, i.e., $(CA + CM_2)/2$, and finding the midpoint of BM_1 is tantamount to taking the average of CB and CM_1, i.e., $(CB + CM_1)/2$. However, we saw above that it is not the arithmetic means that are equal but rather the geometric means of CA and CM_2 and of CB and CM_1. Here is where our method shows the approximate yet extremely close nature of Alberti's method. For line segments such as CA and CM_2, whose difference is very small relative to their large size, the arithmetic and geometric (and harmonic) means are virtually identical.[81] No wonder Alberti's quick method was used by D'Alembert over three centuries later!

Interpretation of Alberti's method has not been without some confusion. Vagnetti praises the exactness of Alberti's method, without providing any analysis of it – and perhaps without really understanding it. Indeed, Vagnetti bases his schematic for Problem 13 (fig. 29) on an illustration from the Bonucci edition (fig. 30), which is at best potentially misleading.

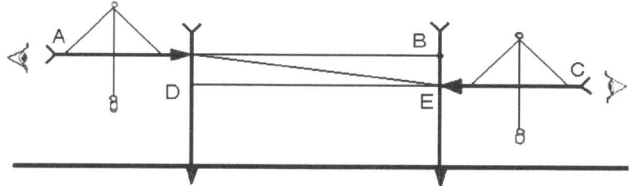

Fig. 29. Illustration of levelling water over large distances from Problem 13 after [Vagnetti 1972: 227]

Fig. 30. Illustration of Problem 13 from Bonucci [Alberti 1844-1850, vol. IV (1847): plate ix]

This figure ignores the curvature of the Earth and, therefore, the fact that the vertical lines in the figure, even if truly vertical at the individual points on the Earth's surface, would not be parallel since they would converge at its centre if extended. (Of course, this

---

[81] Consider that the radius of the Earth is roughly 6.4 million meters, and let us say that the larger of AM_2 and BM_1 is on the order of 10 meters. Then we are working with 6400000 and 6400010 meters; the arithmetic, geometric, and harmonic means of these numbers are 6400005, approximately 6400004.999998, and approximately 6400004.999996, respectively. The difference is on the order of millionths of meters, i.e., microns. If a contrasting case is desired, consider numbers whose difference is not small relative to their sizes, say 1 and 2 meters. The arithmetic mean is 3/2, or 1.5; the geometric mean is the square root of 2, approximately 1.414; and the harmonic mean is 4/3, approximately 1.333. The arithmetic mean and geometric mean are significantly different, off by roughly 9 centimeters.

matters only if the points are very distant, but this is precisely the issue that Alberti is trying to address.) In other words, point B will not simply be at the same height as point A, nor will point D be at the same height as point C. (Indeed, if the corresponding points were at the same height, there would be no need for this lengthy discussion.) It is certainly not clear from the figure how Meliaduse would proceed, given Alberti's words: 'and of the whole difference take the half and this will be suitable measuring for you'. In any case, the figure from Gal. 10, and to a somewhat lesser extent that of Grayson, seem to correspond to Alberti's instructions more faithfully.

The next passage is rather confusing, as the translator's note on the first sentence explains.[82] Alberti discusses here the behaviour of water on a non-level surface, depending on how tilted the surface is. Erosion of river banks is discussed, though again, it is not altogether clear what Alberti is trying to explain to Meliaduse. While some modern readers may find this passage of interest, there is not really any mathematical content for us to consider.

Of greater interest to us is the final use described by Alberti for the *equilibra*, namely to weigh loads that are not too heavy. He has already mentioned the fact that loads hung at either end are determined to be equal by ensuring that the weighted string coincides with the equilibrium point marked on the spear. But now he addresses the situation where unequal loads are hung at either end, and he gives a rule to determine the relative weights of the loads: 'As many times from the end of the spear to line AE goes into the remaining part of the spear, that many times one of these weights goes into the other' (fig. 31).

If one ignores the weight of the spear itself, Alberti's rule is correct, and it involves some nice mathematics. The principle required to demonstrate its correctness is that torque is force times distance, where the tricky thing is that the distance must be measured along a line perpendicular to the direction of force. First consider a scale with a rigid horizontal member of negligible weight, balancing on a point with different weights at the ends, as in fig. 32.

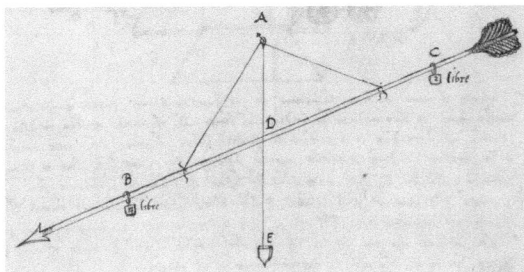

Fig. 31. Third Gal. 10 illustration for Problem 13

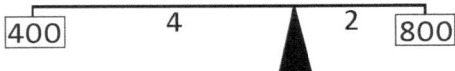

Fig. 32. An illustration of equality of torques

---

[82] See note 148, p. 45 in this present volume.

Suppose 81.6 kilograms, equivalent to about 180 pounds (chosen because it results in a force from gravity of about 800 newtons), is 2 meters from the pivot point, and a second weight, half as heavy, is 4 meters from the pivot point on the opposite side. In order for the system to be in equilibrium, the torques must sum to zero. This means the counter-clockwise torque on the left side must counterbalance the clockwise torque on the right side. The former is 400 newtons times 4 metres, which equals 1600 newton metres, and the latter is 800 newtons times 2 metres, the same magnitude of torque. Note that since the force of gravity is vertical and the rigid member is horizontal, the distances are indeed measured perpendicular to the direction of force, as is required.

A hanging balance (such as Alberti's *equilibra*) works on the same principle, that the torques must counterbalance each other, but these are more difficult to calculate when the rigid member (Alberti's spear) is at an angle to the horizontal. Before we proceed, note that there is a slight labelling error in fig. 31: point *D* seems to signify the intersection of the vertical weighted string with the spear, while Alberti calls *D* the wax that is used to mark the equilibrium point when he first introduces the *equilibra* (see fig. 24). Of course, these would coincide when the *equilibra* is at equilibrium, as in fig. 24, but fig. 31 shows it tilted. As it turns out, though, this labelling error will make our commentary easier from a notational point of view; indeed, for what follows let us view *D* not as the wax but rather as the intersection of the vertical weighted string with the equilibrium point of the spear.

In order to proceed with the mathematical analysis, one would have to resolve line segments *BD* and *CD* into their horizontal and vertical components, because it is the horizontal component that is perpendicular to gravity. More precisely, the pertinent values are the horizontal distances at the pivot point, *A*, which we label as *x* and *y* in fig. 33, but these are the same as the horizontal components emanating from point *D*, labeled *s* and *t*. The horizontal component of *BD*, *s*, equals *BD* multiplied by the cosine of the angle between the spear and the horizontal, labeled α, and similarly for the horizontal component of *CD*, *t*. But we can describe the situation without using trigonometry, and using similar triangles instead.

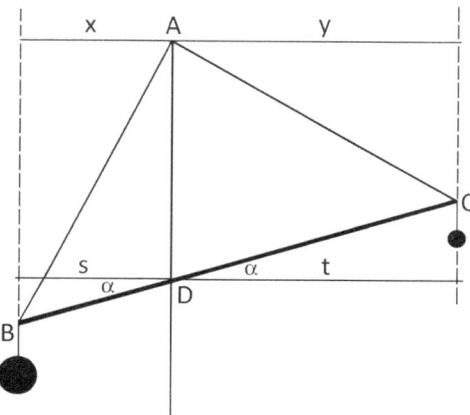

Fig. 33. An illustration of equality of torques, in the case of the *equilibra*

Equality of torques dictates that *s* times the weight on the left is equal to *t* times the weight on the right. One can now perform what could be called 'reverse cross multiplication'[83] to arrive at the proportion,

$$s : t :: \text{the weight on the right} : \text{the weight on the left}.$$

Here is where the similar triangles come into play, for as Alberti would have known,

$$s : t :: BD : CD.$$

Since the left sides of these proportions are the same ratio, *s* : *t*, the ratios on the right sides of the proportions are equal (via *Elements* Book I, Common Notion 1, for example[84]), yielding

$$BD : CD :: \text{the weight on the right} : \text{the weight on the left},$$

which is Alberti's rule.

His verbal statement of the rule, cited above, is a bit vague, but his numerical example makes his intention clear: 'let the spear be six feet long; let there be from end B a weight of 4 pounds, and from end C a weight of 2 pounds; you will find that string [AE] will be 2 feet away from the 4-pound weight, and the other part will be 4 feet'. Note that when he specifies the spear to be six feet long, this would correspond to line segment *BC* in fig. 29 being six feet, not the entire length of the spear.

Leonardo da Vinci directly references the *Ludi* when he registers two complaints about Alberti's description of the mathematics of the *equilibra*:

> Battista Alberti says in a work of his entitled *Ex ludis rerum mathematicarum*, that, when the balance *abc* will have the arms *ba* and *bc* in double proportion, with the weights attached to its extremities positioning it in this way, [the weights] are in the same proportion as are those arms; but it is the other way around, that is the larger weight in the smaller arm. Experience and reason demonstrate that this is a false proposition: because where he places opposing weights 2 against 4 in the balance that itself weighs 6 *libre*, it wants to be 7 against 2; and thus the balance will remain firm with equal resistance in its arms. And here the author erred because he did not mention the weight of the beam of the balance which is unequal in weight.[85]

---

[83] Euclid's two propositions covering cross multiplication – *Elements* VI, Prop. 16 for lines (which is geometric, corresponding to magnitudes) and *Elements* VII, Prop. 19 for numbers (i.e., whole numbers) – are both 'if and only if', so that reverse cross multiplication is also covered; see [Euclid 1956: v. 2, 221ff and 318ff, respectively].

[84] See [Euclid 1956: v. 1, 222]; this is called the 'transitive property' in modern terminology.

[85] Cf. [Pasquale 1992: 46–47]: *Dice Battista Alberti in una sua opera titolata Ex ludis rerum mathematicorum [sic], che, quando la bilancia abc arà le braccia ba e be [sic] in dupla proportione, che ancora li pesi alli sua estremi attaccati che 'n tal modo la dispongono, son nella medesima proportione che sono esse braccia; ma è conversa, cioè il peso maggiore nel braccio minore. Alla qual cosa la sperienza e la ragione dimostra esser falsa propositione: perché dove lui mette li pesi opposti 2 contra 4 nella bilancia che in sè pesa 6 libre, vole essre 7 contro 2; e così resterà la bilancia ferma con equali resistenzia di braccia. E qui errò esso autore per non fare mensione del peso dell'aste della bilancia che è ineguale di peso* (trans. María Celeste Delgado-Librero).

The second complaint is valid, as we will discuss in a moment, but the first complaint seems to be based on a misreading of Alberti's rule, which, although vague, is clarified by Alberti's numerical example, as we stated above. Leonardo seems to interpret Alberti's rule backwards, i.e., that the heavier weight will be further away from the line (*AE*) going through the wax. Of course, Leonardo may have been working with a corrupted *Ludi* manuscript that had this rule backwards, perhaps even in the numerical example. In any case, Leonardo's own numerical example is not altogether clear, but the fact that it involves the weight of the beam itself (Alberti's spear) is more significant. This brings us to Leonardo's second complaint, underscored by the final sentence quoted, which definitely has merit, since the weight of the spear would indeed be a factor in the accuracy of the *equilibra*, especially when it is tilted significantly. In other words, if α is small, then Alberti's rule provides a decent approximation, but for larger angles the weight of the spear becomes more of an issue. To be fair to Alberti, we should note that in our own example above (involving the counterbalancing torques of 1600 newton meters), we also assumed that the horizontal member has 'negligible weight' for convenience. In terms of Alberti's *equilibra*, the heavier the objects being weighed are in relation to the weight of the spear itself, the more negligible the spear's weight is.

Alberti next deals with weights of very large loads, and then he revisits the *equilibra* in Problem 15.

Problem 14
Measuring the weight of heavy loads (Gal. 10 fols.11r-11v)

It is useful to draw on the final book of Vitruvius's treatise for some background pertinent to this game.[86] In Book X, chapter ii, Vitruvius describes hoisting machines, cranes that incorporate pulley systems with multiple pulleys arranged in two sets of pulley-blocks, an upper pulley-block and a lower one, which divide loads significantly by affording many back-and-forth turns between the pulleys. He names these cranes based on how many total pulleys there are, e.g., when 'there are two pulleys revolving in the lower pulley-block and three in the upper, the machine is called a *pentaspastos*'.[87] In chapter iii Vitruvius discusses types of motion, starting in section 1 with an astute observation on how much is gained by marrying circular motion with motion along a line. The application that Vitruvius is discussing here is raising large loads, but to this day so much of our machinery is based on some combination of linear and circular motion. In sections 2 and 3 he rehearses the basics of the pulley, lever and fulcrum. This leads to a discussion in section 4 on a steelyard (or steelyard scale, or Roman scale), a balance used by the ancient Romans (fig. 34).[88] The principal feature of the steelyard is that the arm on the side of the calibrated scale and weight (the left side of the steelyard in fig. 34) is longer than the arm from which the object hangs, allowing for relatively large loads to be weighed with significantly smaller weights.

---

[86] See [Vitruvius 2009: 276ff].
[87] [Vitruvius 2009: 280–281 (X, ii, 3)].
[88] Souffrin uses the term *balance romaine*; see [Alberti 2002: 96].

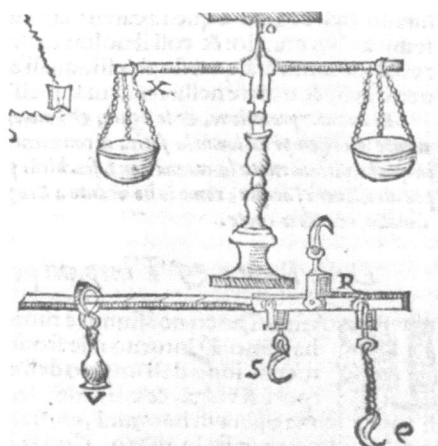

Fig. 34. A detail from the Barbaro translation and commentary of [Vitruvius 1584: 459]; the lower scale is a steelyard

The game is to weigh a very large load, e.g., a cart full of goods together with the oxen, using a steelyard having a fifty-pound weight. Of course, this calls for significant infrastructure to be built up around the steelyard, in order to produce the mechanical advantage necessary to pull off the feat. Indeed, this game involves the largest construction project in the *Ludi*, by far: Meliaduse would need to '[o]rder a bridge similar to one of these drawbridges' as just the first step. Add to this a huge beam built into the drawbridge housing, balanced at the housing and extended beyond the housing on both the drawbridge side and the other side of the door. (The figures are problematic in both Gal. 10 and Grayson, since in Alberti's construction line segment CB should be longer than AB, which would of course result in a division of load.) The next main ingredient is a pulley at point B – or, more likely, a system of pulleys between points B and D, as discussed below. Finally there is the steelyard at point E, connected in an interesting way. The place on the steelyard from which it would normally hang becomes the place where the cable from B is attached, and the place where the object to be weighed usually hangs is instead attached to the ground.

All of these simple machine components connected together act to divide the weight down by a factor on the order of 200, as reasoned by Souffin.[89] He makes the case that this game remained a mystery to Grayson and Vagnetti, and then he provides a convincing analysis to understand it properly. The main point of confusion is the single pulley vs. multiple pulley system, which occurs between points B and D in the schematic by Vagnetti. (In the Gal. 10 illustration, fig. 35, D is apparently labelling the general level of the bridge/floor/ground opposite the drawbridge, which is actually reasonable given the two appearances of D in the text, but Vagnetti's positioning of D, fig. 36, is more useful for our discussion.) Alberti's initial description of the line from B to D seems to describe a single pulley wheel only, but when he discusses the mathematics of the process he

---

[89] See also [Alberti 2002: 97]; we shall cite a good bit of the analysis, from pages 96–99.

indicates that there are plural turns of the cable, as indicated by Souffrin. This would suggest a pulley system with two pulley-blocks and multiple turns back and forth, as Vitruvius describes, which is sometimes called a tackle.

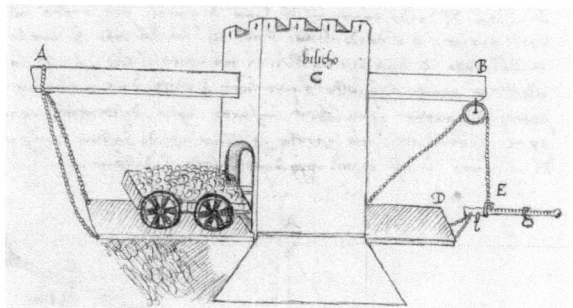

Fig. 35. Gal. 10 illustration for Problem 14

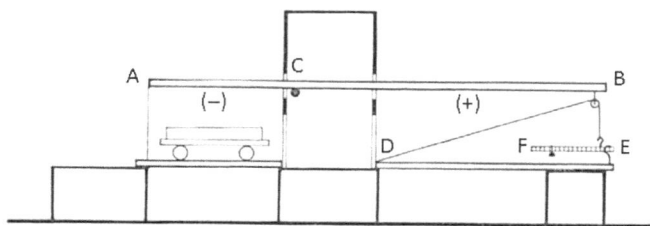

Fig. 36. Illustration of Problem 14 after [Vagnetti 1972: 230]

Souffrin suggests two reasons for the confusion regarding single pulley vs. tackle. Firstly, the illustrations indicate only one cable between points B and D; note in fig. 36 that Vagnetti's schematic depicts a single line. Secondly, Alberti's use of the word *tagliuola* may have been misunderstood, as Souffrin discusses at some length. There are many possible variants and meanings to *tagliuola*, but Souffrin argues that some sort of pulley system is indicated. There is one more piece of the description that has the potential to cause a bit of confusion, the 'spool' at point D, but this is the shaft of a winch.[90] If we accept *tagliuola* to mean a set of pulleys forming a tackle, then it would indeed need to be tied off tightly at point D, as would be accomplished by a winch.

Alberti suggests determining a conversion factor between the weight(s) at the steelyard and the weight of the load: '[once] you count how many pounds of the cart move an ounce of your [little] beam scale, by that rule you can then [always] weigh all the others'. For this to be an accurate comparison from load to load, each one needs to be placed at the same place on the drawbridge platform. In fact, Souffrin states that there are four divisors of load, the additional one (besides the huge beam, the tackle, and the steelyard) being the drawbridge, which is itself a lever. In any case, for Alberti's determination of a conversion factor to be valid, the position of the centre of mass of each load must be the same.

---

[90] See [Mancini 1917: 201]. Souffrin also describes the existence between B and D of a winch (*touret*) together with the pulley (*moufle*) [Alberti 2002: 96].

Problem 15
Using an *equilibra* to aim a bombard (Gal. 10 fols. 11v-12r)

While many of the games could be used during wartime in specific situations, this is the only one that is specifically about a military application of mathematics.[91] Alberti does couch his language regarding the application: 'It seems to me that it should not be omitted, more to show you a use for your scale than to reason about things alien to your dignity and authority'. It is not surprising that at least one military application made it into the *Ludi*, given that Vitruvius himself listed several. These all occur at the end of the first-century Roman treatise, chapters x through xvi of Book X, and the contraptions are quite involved.[92] The sentence before, i.e., the last sentence of chapter ix, is important enough in terms of *Ludi* to quote from two different translations of Vitruvius:

> Schofield: 'I have now described carefully how machines that are of practical use and sources of amusement should be constructed in peaceful and tranquil times'.[93]

> Morgan: 'I have described how to make things that may be provided for use and amusement in times that are peaceful and without fear'.[94]

On the one hand, this sentence is simply a transition that Vitruvius makes to lead into the wartime applications. On the other hand, there is a key word that is common to both translations, 'amusement', from the Latin *delectationem*, which correlates nicely with what Alberti seems to envision with the *Ludi*.

The game is to refine the aim of a bombard, a medieval cannon usually loaded with a large stone, by recording the angle and direction of each fire (as well as the amount of powder and the size stone used) and adjusting accordingly. The unwritten issue can be realized by thinking about the kick-back that a cannon undergoes when it is fired. The bombard, a rudimentary cannon, must have moved quite a bit upon being fired. So one needed to be able to recover the prior placement and adjust to a new placement. To record the angle before firing, Alberti has the bombardiers suspend (presumably just by holding it) the *equilibra* (constructed in Problem 13) directly over the bombard, aligned with notches made on the rims at either end. The basics are clear: record the position of the bombard relative to the *equilibra*, and that of the *equilibra* relative to a point marked on the ground behind the bombard, so that after firing, adjustments for aim can be made, and then the *equilibra* can be repositioned and the bombard aligned under it.

---

[91] In the next problem, Problem 16, Alberti does mention the possibility of military use, so perhaps this is the reason he placed them consecutively.

[92] The illustrations in [Vitruvius 2009] are chosen well, and the ones for the military machines taken after Fleury's edition of Vitruvius are very helpful for understanding his descriptions. For his primary plates Richard Schofield chose the illustrations from Daniele Barbaro's edition of Vitruvius, including, of course, many drawn by Palladio, which certainly deserve the wider distribution they are now receiving.

[93] [Vitruvius 2009: 303 (X, ix, 7)].

[94] [Vitruvius 1960: 303 (X, ix, 7)].

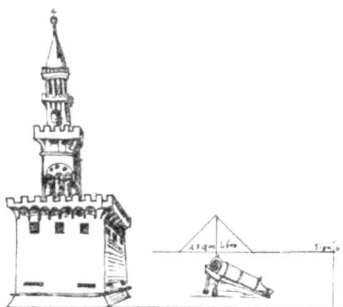

Fig. 37. Grayson illustration for Problem 15 (from *Ludi* manuscript Riccardiana 2942,1)

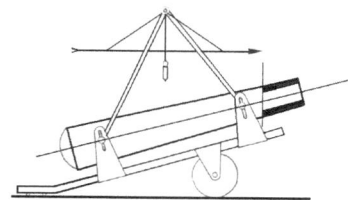

Fig. 38. Illustration of Problem 15 after [Vagnetti 1972: 234]

But the specific instructions are not easy to follow completely. The first question is, should the *equilibra* be level, as shown in the illustrations from Gal. 10, Grayson (fig. 37), and Vagnetti (fig. 38), or should it be tilted at the same angle as the bombard? Alberti's instruction to 'note where the weighted line strikes the *equilibra*' seems to indicate that the *equilibra* would be tilted, for otherwise the weighted line would clearly always strike the *equilibra* in the middle, at equilibrium. Another issue of confusion is the 'mark' (*segno*):

And along the straight line of the end where [the *equilibra*] is placed, aim at the contrary place opposite the place where you want to hit, and where the aim of your *equilibra* strikes, put a mark.

It is clear that the mark is supposed to be placed to the rear of the bombard and that it should be obtained by sighting back along the scale. But if the *equilibra* is level, the mark would be on a vertical line or spear, as in the Grayson illustration (labelled *signo*), whereas if the *equilibra* is tilted, the mark would be on the ground.

In either case, the most confusing instruction is the final statement concerning readjusting the bombard after it is fired:

You will see where it hits, and adjust the high and the low and the slope the second time moving the mark that you had placed behind, and to that mark so moved aligning your *equilibra*, and moving the bombard under the *equilibra*. It should be that this mark is as far away as the place you want to hit.

Why should the mark be as far away as the target? And if it is – if this is some sort of ideal location for the mark – why would it ever be adjusted?[95] The wording, at least as it has come down to us, is confusing, and in any case it may be difficult to achieve high accuracy with this method. However, Alberti's point is certainly valid: if you record the position of the bombard before shooting, then you can learn from your misses and adjust accordingly.

Problem 16
Constructing a circular instrument for measuring angles in order to draw the map of a city and to estimate distances (Gal. 10 fols. 12r-14r)

The primary tool of this game is a simplified version of the instrument more fully described by Alberti in his treatise *Descriptio urbis Romae*, translated by Carpo and Furlan (and Peter Hicks) as *Delineation of the City of Rome*, and which Grayson dates to the 1440s.[96] The instrument is a circular device, 'at least a *braccio* wide' (a fairly large size, conceptually an arm's length), having its circumference divided into many equal parts, 'the more the better'; Alberti indicates that his usual choice is 48 'degrees' each of which has 4 'minutes' for a total of 196 minutes.[97] Of course, Alberti's degree and minute differ from the modern standards (360 degrees in a full circle, 60 minutes in a degree), but the concept is the same. The device is used as an aid to determine angles: the user, located at point A, orients the zero-degree mark of the circle in a certain direction, say B, then sights to a point C, so that this line of sight passes through the centre of the circle, and thereby reads off angle BAC. This angle can be useful in and of itself, but it can also be paired with the distance from A to C, in order to produce an example of what we now call polar coordinates (think of A as the origin).

Indeed, Alberti's *Descriptio urbis Romae* provides a 'map' of Rome consisting of {angle, distance} data points of the major features of Rome at his time (he chose the Capitoline Hill as the origin).[98] The circular device described in the *Descriptio urbis Romae* is meant solely as an aid in plotting the points, already recorded by Alberti, on paper, in order to produce an actual map. As such, the device also includes a rotating radial

---

[95] It may instructive to consider how Alberti starts this game: 'Recall that another time I explained how it is possible to aim a bombard without seeing where the stone has to go.' Perhaps Alberti is describing a practice round with the bombard, where Meliaduse is supposed to aim at a mark that is 'as far away as the place you want to hit'. Another possibility is that somehow similar triangles are to be used between the first and second shots of the bombard (whether this is a practice round or not), by maintaining the same angle of the *equilibra*. However, adjusting 'the high and the low and the slope' would seem to indicate changing the angle.

[96] Cf. [Alberti 2007] and [Grayson 1998: 299]. As mentioned in the Introduction to this present volume, the *Descriptio urbis Romae* is the one mathematical treatise of Alberti that we do not include here, because Carpo and Furlan, with their 2007 publication, have very recently made an excellent English translation and commentary available.

[97] For a discussion of 48-fold division, see Lionel March's commentary to *On Writing in Ciphers*, p. 192 in this present volume.

[98] As discussed above in the introduction to this commentary, the reason that Alberti did not simply provide a map in illustration form is that the reproductions possible with the manuscript tradition were limited in accuracy by the ability of the copyists. The inevitable result is a worsening of accuracy as copies upon copies are made.

arm, with one end at the centre of the circle, revolving about it, and the arm coincides with the circle's radius at any turn. The scale that Alberti gives for the radial arm (with a maximum value of fifty) makes it easy to work with the distance data he lists (the largest distance is 46), so that the points given in the {angle, distance} pairs from Rome could be plotted to scale on any piece of paper at least as large as the circular device.

There is a similar device described by Alberti in *De statua*, the difference being that the radial arm extends past the circumference of the circle. The circle is centred at the top of the head, since '[t]he main theme of *De statua* is the measurement of men or statues of men, with the object of reproducing their exact form'.[99] Alberti got a good deal of play out of the idea for this device, simply because polar coordinates are so very useful in various situations. As for whose idea it was, we may never know. Alberti may have invented the circular device, as Mancini states, although Vagnetti points out that an anonymous Viennese cartographer suggested an analogous method in 1432.[100]

What is quite remarkable about the use of the circular device in the *Ludi*, from a mathematical point of view, is that a radial arm is not needed, i.e., distances are not necessary. This is never explicitly stated by Alberti, but it is implicit through the fact that his instructions for Meliaduse do not mention distances at all. Moreover, it will become apparent through careful analysis that only angles are needed for this game.

At the beginning of Problem 16, Alberti introduces the circular instrument as something that Meliaduse may find useful in wartime. He then mentions its use in his depiction of Rome, another case, as in the mention of De re aedificatoria in Problem 13, where such information allows historians to determine the chronology of Alberti's writings. He gets started in earnest with the description of the instrument itself (fig. 39), which we already discussed above. Note that the Gal. 10 illustration of the instrument has 32 degrees rather than 48, and it is quite a rough drawing.

The second figure for this problem, having three towers, with a circle above each and a triangle connecting the centres of the circles, illustrates what Alberti describes next (fig. 40). As one would gather from the figure, Alberti instructs Meliaduse to go to the top of each tower (or other high place) and record the angles 'of all you see' (including the other towers, of course). The method he suggests to make the sightings accurately is to hold up a string with plumb bob lined up with the line of sight and the centre of the circular device. Notice that he does not mention taking any readings of distances, neither between towers nor from the towers to any of the other landmarks.

The next step is to produce a *pictura* from the data collected at the towers. Before continuing, let us consider what Alberti may have meant by *pictura*. Translating *pictura* as 'painting' in this problem seems problematic, since what will be produced by Alberti's method is a plan view, a view from above, indeed a map. In contrast, Alberti conceived of a painting as a proper scene painted in perspective, as we know from *De pictura* and *Elementa picturae*.[101] Souffrin translates one instance of *pittura* (this is how it appears in

---

[99] [Grayson 1998: 304].

[100] Cf. [Mancini 1882: 113] and [Vagnetti 1972: 238–240]; Vagnetti states that Alberti was at least the first to use polar coordinates to survey an entire piece of land.

[101] See our commentary to *Elements of Painting* for an expansion of this analysis, p. 153 in this present volume.

Grayson) into French as *plan* (plan, map) and a second instance as *figure* (figure).[102] Analogously, we choose the more general 'picture' rather than 'painting' for *pictura* in this problem.

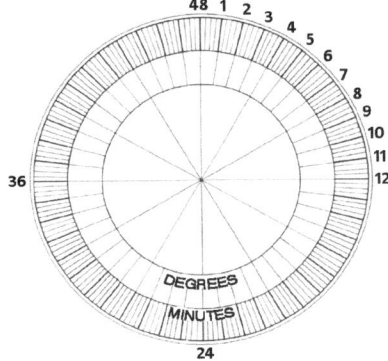

Fig. 39. Illustration of the circular instrument of Problem 16 after [Vagnetti 1972: 236]

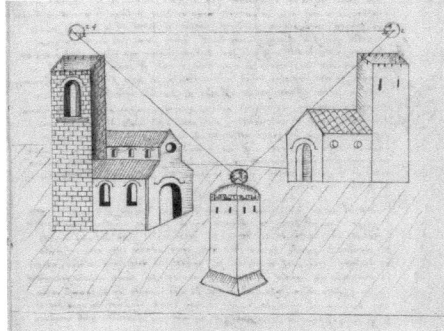

Fig. 40. Second Gal. 10 illustration for Problem 16

Proceeding now to Alberti's instructions for Meliaduse for producing the picture on paper at his drawing table from the information collected *in situ*, we carefully consider the mathematics in order to fully appreciate the cleverness of this game. First Meliaduse is to pick a point on the paper to correspond to the tower at the castle (and label it 'Castle'), its position chosen by planning ahead and envisioning how the 'whole picture' will be laid out. Alberti then instructs him to make a small paper version of the circular instrument that was used to collect the data, centre it at the 'Castle' point, and draw lines emanating from this point that correspond to all the angles measured from the castle tower. Next, Meliaduse must pick a point on the specific line from the 'Castle' point that corresponds to

---

[102] Cf. [Alberti 2002: 66]: *Vous commencerez par marquer, sur la surface sur laquelle vous voulez faire votre* plan, *un point à partir duquel vous pensez pouvoir dessiner toute la* figure,... (emphasis ours).

one of the other two places where you aimed at the things, and on this second point place also another similar small paper instrument, and set it so that it corresponds to the line of the number which is called in your notes 'Castle' that is, that both instruments are on a line together corresponding to each other....

Having placed this second point corresponding to a second tower (which is not named), Meliaduse is then to draw all the lines radiating from this point that correspond to all the angles measured from the second tower. Where the lines from the 'Castle' point intersect the correspondingly named lines from the second point, Meliaduse is then able to determine the positions of all the landmarks measured *in situ* (to be precise, all the landmarks that were measured from both the castle tower and the second tower). Alberti gives an example of such a landmark, named Santo Domenico, which, it should be noted, is not to be confused with the third tower. Alberti has not yet mentioned the third tower in this reproduction process, but he will soon, as we shall discuss in a moment.

Before proceeding, however, we must admit that we wondered at first, as Meliaduse may have, how this would work without incorporating knowledge about the distance between the castle tower and the second tower, not to mention other distances, but this is the beauty of Alberti's method. As long as all the angles at these two points are correctly measured *in situ* and faithfully reproduced on paper – which may be difficult in practice but is possible in theory – the method works precisely. To see this, consider two triangles, one having vertices at the castle tower, the second (unnamed) tower, and Santo Domenico, and the other being the analogous smaller triangle on paper, i.e., having vertices at the 'Castle' point, the second point, and the 'Santo Domenico' point. Assuming correct measurement and reproduction of the angles mentioned, i.e., assuming that the angle at the castle tower equals the angle at the 'Castle' point and the angle at the second tower equals the angle at the second point, then since the third angle of any triangle is determined if the first two are known, the angle at Santo Domenico must equal the angle at the 'Santo Domenico' point.[103] Thus, the two triangles – the one in real life and the one on paper – are similar triangles, so that they are scaled versions of each other. This is precisely what is required to draw a picture (or a map) to scale.

In theory, then, all one would need is to have measured the angles carefully from two towers, but Alberti then explains how Meliaduse can use the data collected from the third tower:

> If it happens that two of these lines do not cut well together so that their angle is not very clear, place another similar small instrument on the third point where you noted the things, and set this similarly to the others so that their lines correspond to each other, and this will show you everything in full.

Actually, one wonders how Meliaduse would have any way of knowing whether corresponding lines from the 'Castle' point and the second point 'cut well together' (intersect well) without considering the data from the third point. Perhaps a point of intersection would be clearly wrong if, for example, a simple writing error was made in recording the angle at one of the two towers. In any case, what Alberti is indicating is the

---

[103] The elementary fact that two angles of a triangle determine the third is contained in *Elements* I, Prop. 32 [Euclid 1956: v. 1, 316 ff].

fact that the data from the third point can be used to double-check the intersections from the first two points. This could certainly be useful, since presumably there would be some human error involved in determining the angles *in situ* and, albeit to a lesser extent, reproducing them on paper. All in all, this is quite a game!

After explaining one application that he himself made of this method,[104] Alberti ends the game by offering a way to measure distances. This should not be surprising, given that implicit to this method is the use of similar triangles, which are employed by Alberti in Problems 1 through 7 to find lengths of towers, widths of rivers, and depths of wells. He now makes the similar triangles explicit, the larger one life-size and the smaller one drawn according to the above method. One side of the life-size triangle is a long corridor at the castle; its other two sides are the lines of sight from the endpoints of this corridor to a nearby tower, the Torre dell'Asinello. Using the aforementioned circular device, Meliaduse can measure the angles of this life-size triangle at the endpoints of the corridor, which is all he needs to be able to draw a smaller depiction of it somewhere on the floor of the corridor. He can measure the length of the corridor, and he wants to know the distance to the tower (from one of the endpoints of the corridor). He can measure the corresponding two sides of the smaller triangle on the floor, and since corresponding sides of similar triangles are in proportion (*Elements* VI, Prop. 4), he can routinely solve the proportion to find the distance sought.

Problem 17
Measuring the distance to a remote but visible place (Gal. 10 fols. 14r-14v)

Problem 17 is certainly the most confusing of the *Ludi*, and it clearly has been received by us in corrupted form. It is also the most infamous game, at least by measure of how much analysis on it has been generated over the past decade or so. All known manuscripts are corrupted at a key point of the exercise, and the illustrations have fared no better. Fortunately the received texts have been carefully studied, from both a philological and mathematical point of view, so that we can be quite confident in what the content is supposed to be. Much of the analysis is due to Francesco Furlan and Pierre Souffrin, two of the preeminent Alberti scholars of recent times, and we will be drawing from [Furlan 2001], [Furlan and Souffrin 2001], and [Alberti 2002] in this commentary.[105] Unfortunately, however, even given the reconstruction of Alberti's method provided by Furlan and Souffrin, it works only in certain cases, and not, as Alberti states, for 'any large distance' – indeed, not even in the case he provides in this game. That is to say, it seems that he was guilty of some faulty logic, as we shall see below.

Alberti's method seems to be of his own invention.[106] The game is to determine the distance from the viewer's location to a remote but visible place. The example chosen by Alberti is the distance to Bologna from a location near Ferrara, a distance of over 20

---

[104] See note 194, p. 57 in the present volume.

[105] The introduction of [Alberti 2007] contains some additional commentary, as well; in particular, see the figures on p. 39, which are also given in [Furlan and Souffrin 2001: 20].

[106] See [Souffrin 1998: 90].

kilometres.[107] The modern reader may first wonder if it would be possible to see Bologna from such a distance. Of course, the distance alone is not the issue: after all, we can see stars in the sky that are trillions of kilometres away. The potential obstacle to seeing over large distances on the surface of the Earth, in general, is the presence of obstructions to the line of sight, including haze in the air, possibly exacerbated by the Earth's curvature (with which Alberti already contends, in the context of levelling water; see Problem 13). So let us assume that Meliaduse would have a location from which a certain landmark in Bologna is viewable, in order for Alberti's method to be plausible.

Alberti first instructs Meliaduse to find a large field from which Bologna can be sighted, and to mark with spears two points in the field at a distance of a thousand feet or more, 'as long as one sees the other and each sees Bologna, so that between the three of them, that is, Bologna and the two spears, they make a well partitioned triangle'. Just how 'well partitioned' this triangle is will be the crux of the matter, as we shall see below, but let us continue. These spears are labelled A and D by Alberti, as in the illustration from Grayson (fig. 41; the Gal. 10 illustration is not worth considering in this game, as will be understood after reading our commentary).

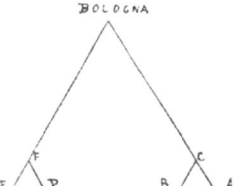

Fig. 41. Grayson illustration for Problem 17 (from *Ludi* manuscript Riccardiana 2942,1)

Now Meliaduse must sight from spear A to spear D, and on the line of sight mark a point B (with a cherry) twenty feet away from spear A. Then he must sight from spear A to Bologna, and on this line of sight mark a point C (with a rose) thirty feet away from spear A. Having created the triangle ABC on the ground, Meliaduse must measure the three sides precisely and record them, because Alberti now instructs him to make an exact copy of this triangle at point D, based on how Furlan and Souffrin have reconstructed the next two steps. Here is our English translation of the two infamous sentences in question from Gal. 10 (with the bold text borrowed from Grayson, and our insertion in angle brackets):

This done, move to the second spear, [**and turn your face towards Ferrara, and move away twenty-five feet, and along this second spear**] aim in a straight line at the first spear, and along this straight line made by your <line of> sight, place a cherry as near to this spear exactly as much as B was near to A. Then turn your face towards Bologna, and along the straight line of this spear aim at Bologna, and on the ground on that line place a rose that is as far

---

[107] Furlan [2001: 152, note 13] instructs us that the location of what is called 'your monastery' (*dal monasterio vostro*) in Gal. 10 is actually the village of Monestirolo, near Gaibana, no more than 15 km from Ferrara. While this is basically south of Ferrara and therefore closer to Bologna than is Ferrara, it is still more than 20 km from Bologna.

from the first spear as C was from A in the first triangle, and take a cord from this spear up to the rose.

Regarding the first sentence, it is not immediately clear from the words, certainly not from Gal. 10 alone, that the second cherry is to be positioned not between the two spears but rather past spear D, i.e., so that point D corresponds to point A, and point E (the second cherry) corresponds to point B; however, this is how it is pictured and how it must be interpreted (see the Grayson figure shown in fig. 41, which, although problematic as we shall momentarily discuss, does show the basic position of the copy of triangle ABC to the left of point D). Thus, it should be noted that the so-called 'well partitioned triangle' (which as we mentioned about, is the crux of the matter) does not have the second spear as a vertex but rather point E.[108]

We now face the most problematic sentence of Problem 17, if not all of the *Ludi*, the second sentence cited above, 'Then turn your face towards Bologna...'. In order to address this sentence, it will be helpful to introduce an alternative version of the figure for this exercise (fig. 42).[109]

In this figure, ignore for the moment the line segment extending from E to Bologna, which has not yet been described by Alberti. Instead, focus on the fact that the point marked by the second rose (Rosa) is not the same as the point F, the latter of which has not yet been found. That is, the triangle with vertices D, E (the second cherry), and Rosa (the second rose) is the one that Alberti instructs Meliaduse to construct congruent to triangle ABC – though this is not at all apparent from the second sentence.

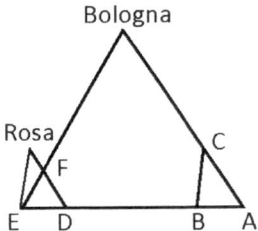

Fig. 42. An illustration for Problem 17 based on the Furlan and Souffrin reconstruction, after [Alberti 2002: 109]

We now turn to the analysis of Furlan and Souffrin, who reconstruct this sentence using a combination of philology, mathematics, and history of science. Furlan [2001] decries the eclectic approach of Grayson in his edition of the *Ludi*, in which an attempt was made to arrive at a correct rendition of the games by combining elements from various manuscripts. To make matters worse, Grayson did not indicate what he had taken from which

---

[108] We do not mean to imply that Alberti was in error, since how 'well partitioned' the triangle is (or is not) would depend not on whether point D or point E is the third vertex, but rather on the distance between the two spears relative to the distance from Meliaduse's monastery to Bologna. We simply mean to point out that Alberti is a bit misleading in mentioning the triangle between the two spears and Bologna in the first place, since this triangle itself is not part of the construction.

[109] This figure is found in [Furlan 2001: 165], [Furlan and Souffrin 2001: 17], [Alberti 2002: 109], and [Alberti 2007: 39], in nearly identical forms.

manuscript, let alone how he had made such decisions. In the case of Problem 17, moreover, Grayson failed even to produce a version that makes logical sense in the present context: the construction of the second small triangle congruent to triangle *ABC*.

To remedy this situation, [Furlan 2001] uses a rigorous philological approach to construct an impressive critical apparatus, including the formation of a tree of potential relationships among the existing manuscripts of the *Ludi*. He combines this with a mathematical analysis of the situation, also detailed in [Furlan and Souffrin 2001], in order to reconstruct the notoriously problematic second sentence cited above. It should be noted in particular that the reconstruction is heavily informed by what Alberti states in the rest of the problem, i.e., the mathematics that he eventually performs, which we shall analyze below. Indeed, Furlan states that the passage in question is 'clarified beyond all reasonable doubt by the reconstruction that Souffrin has been able to work out on an exquisitely geometrical plane, and which can then be precisely determined on a formal and linguistic plane...'.[110]

Our translation of the infamous sentence, based on their reconstruction, is then as follows:

> Then turn your face towards Bologna, and on the ground place a rose distant from the cherry as much as C is from B, and distant from the spear just as much as in the first triangle C was distant from A, and take a cord from this spear up to the rose.[111]

Referring to fig. 42, the instructions given in this reconstruction can easily be seen to reproduce triangle *ABC*, resulting in the triangle with vertices *D*, *E* (the second cherry), and Rosa (the second rose), as desired. Now the rest of Alberti's instructions make sense. Meliaduse must go back to *E*, sight to Bologna, and mark where this line of sight intersects the cord with a stick. Alberti labels the stick *F*, as shown in fig. 42.

The result of the construction is a pair of similar triangles, triangle *DEF* and the triangle with vertices *A*, *E*, and Bologna. Indeed, angle *E* is shared by these two triangles, angle *A* is equal to angle *D* by virtue of the above reconstructed method, and hence the angle at *F* must equal the angle at Bologna. Then by using the fact corresponding to Euclid's *Elements* VI, Prop. 4, as Alberti has done repeatedly (see Problems 1 through 7, as well as 16), he is able to instruct Meliaduse that $ED : EF :: EA : E\text{-Bologna}$. Since the

---

[110] Cf. [Furlan 2001: 162]: *Cosa dovesse nella sostanza contenere il breve passo in tal modo omesso è chiarito al di là d'ogni ragionevole dubbio dalla ricostruzione che su un piano squisitamente geometrico ha potuto operare Pierre Souffrin, ed è poi precisamente determinabile, sul piano formale e della lingua, sulla base delle osservazioni fatte sin qui* (trans. María Celeste Delgado-Librero).

[111] Cf. [Furlan 2001: 165]: *Poi volgete il viso verso Bologna, e in terra ponete una rosa distante dalla ciriegia quanto C da B, e distante dal dardo proprio quanto fu nel primo triangulo distante C da A, e tirate un filo da questo dardo fino alla rosa*; and cf. Souffrin's French version [Alberti 2002: 71, 108]: *Tournez-vous vers Bologne, et placez sur le sol une rose à la même distance de la cerise que C est distant de B, et à la même distance de la flèche que C est distant de A dans le premier triangle, et tendez un fil de cette flèche à la rose* (trans. Stephen R. Wassell, using 'spear' for *dardo* [Souffrin's *flèche*] as we do throughout the translation of the *Ludi*).

first three quantities are measurable by Meliaduse, the fourth, i.e., the distance to Bologna, can be determined.

After stating as much in his 2001 article, Furlan concludes with the following final sentence as a sort of punchline:

> In keeping with the promise in the dedication to Meliaduse, the solution to problem XVII of Alberti's *Ludi rerum mathematicarum* thus turns out to be no less 'practicable' and 'useful' than 'amusing'.[112]

Unfortunately, however, on this point we do not agree with Furlan: the procedure as formulated by Alberti, accepting the reconstruction informed by Furlan and Souffrin, is definitely not practicable. To understand this, note that for the construction (as reconstructed) to work, angle B in triangle ABC has to exceed angle E in triangle AE-Bologna. We will first show that this is not the case with the numbers Alberti uses at the beginning of the game, and then we will comment on the potential for making this exercise useful.

Consider the following three facts, each of which is supported by a correspond figure.

1. If angle B in triangle ABC equals angle E in triangle AE-Bologna, then the ratio of AB to AC equals the ratio of AE to A-Bologna; the converse is also true. This case is shown in fig. 41.

2. If angle B in triangle ABC is greater than angle E in triangle AE-Bologna, then the ratio of AB to AC is smaller than the ratio of AE to A-Bologna; the converse is also true. This case is shown in fig. 42.

3. If angle B in triangle ABC is smaller than angle E in triangle AE-Bologna, then the ratio of AB to AC is greater than the ratio of AE to A-Bologna; the converse is also true. This case is shown in fig. 43.

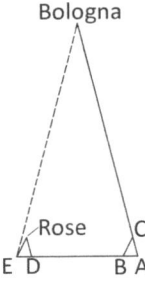

Fig. 43. An illustration for Problem 17 showing Case 3

Before proceeding with the discussion of Alberti's method, let us justify these three facts in their own right. Case 1 is clearly true, because this is the case in which triangles ABC and AE-Bologna are similar. Indeed, the first if-then statement in Case 1 is equivalent to *Elements* VI, Prop. 4, and the converse is equivalent to *Elements* VI, Prop. 5;

---

[112] Cf. [Furlan 2001: 165]: *Secondo il voto dell'accompagnatoria a Meliaduso, la soluzione del problema XVII degli albertiani* Ludi rerum mathematicarum *torna così ad essere non meno «practicabile» e «adoperabile» che «iocunda»* (trans. María Celeste Delgado-Librero).

note that *Elements* VI, Prop. 2 relates to Case 1 as well.[113] As for Cases 2 and 3, they are obvious from the figures indicated, when contrasted to Case 1.

Bologna

Now, for Alberti's construction to work, given the analysis of Furlan and Souffrin, we would need to be in Case 2. But we must be in Case 3, since the length *A*-Bologna is by far the largest distance, and in particular larger relative to *AE* than *AC* is to *AB*, the last two being 30 feet and 20 feet, respectively, as instructed by Alberti. Let us consider these values in some detail. Length *AE* is the sum of *AD* and *DE*, which in turn is the sum of 'a thousand feet with as much more as you wish' and about 20 feet (the latter being whatever the exact figure length *AB* turns out to be). Thus *AE* will be quite a bit shorter than either *A*-Bologna or *E*-Bologna, each of which are over 20 kilometres, as discussed at the beginning of this commentary. To err on the side of making Problem 17 as operable as possible, let us use 20 km as the distance from Monestirolo to Bologna. In order to compare this distance to *AE*, we need to use common units. Unfortunately, there was no standard foot during the time of the Italian Renaissance, and in fact the modern reader might be surprised to learn just how much it could vary: values as low as 22.3 cm and as high as 64.9 cm were used in various contexts and in various regions.[114] Extreme values are fairly rare, however, and most values are somewhere in the range between 30 and 40 cm.[115] Accepting this as reasonable, let us again err on the side of making Problem 17 as operable as possible and choose 40 cm for Alberti's foot; to balance this, it would be reasonable to ignore the 20 feet given by *DE*, which is fairly negligible anyway, and consider *AE* to be 1000 feet, equivalent to 400 metres (based on our above generous conversion factor). Dividing this into 20 km, we conclude that the ratio of *AE* to *A*-Bologna will be approximately 1:50, certainly much smaller than the ratio of *AB* to *AC*, which is 20:30 or 2:3 by construction. Thus, Case 3 above is clearly the correct case, and Alberti's method simply will not work. We show in fig. 44 the overall triangle *AE*-Bologna, for the reader to visualize just how far from 'well partitioned' it is!

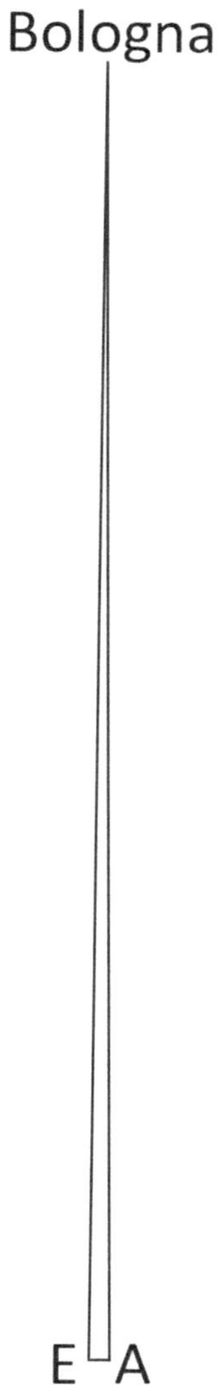

E A

Fig. 44. A scale depiction of the overall triangle of Problem 17

---

[113] See [Euclid 1956: v. 2, 200ff and 194ff];

[114] See [Zupko 1981: 195–200].

[115] Of interest may be the determination of the *piede* used by Palladio at the Villa Cornaro based on a thorough survey of the building, found to be 34.8 cm to the nearest millimetre, a seemingly preposterous level of accuracy that is nonetheless supported by two independent mathematical methods [Mitrović and Wassell 2006: 41ff].

Alberti had the right idea, of course, to make AC (30) longer than AB (20). One could extend AC to make the method work, but it would need to be on the order of 50 times 20, or 1000 feet, which may (or may not) have been possible in the field Meliaduse would have used. One could instead decrease AB, or perhaps both increase AC and decrease AB. Changing 20 feet to 1 foot and 30 feet to 50 feet, for example, would be on the right order of magnitude, but then one can well imagine that any error in measuring these distances, and in copying the triangle ABC, would be quite significant indeed when the proportion is solved at the end of the process.

Alberti also had the right idea to make AD quite large, and he does specify that this distance should be 'a thousand feet and as much more as you wish,' just so long as all of the sightings are possible and the overall triangle (in reality, meaning triangle AE-Bologna) is a 'well partitioned triangle'. Here we return to the crux of the matter. What is a well partitioned triangle, anyway? Something close to equilateral, as is shown in most of the illustrations? Yes, an equilateral triangle is fine for a schematic, but it can also be misleading. The fact is that triangle AE-Bologna would be anything but well partitioned, unless AD was on the order of a few kilometres, at least (the actual length necessary would depend on the definition of well partitioned, of course). So while the method might work on paper, with a schematic triangle close to equilateral, it will not work in the field, at least not without significant adjustment of the numbers. Let us reconsider adjusting AD, then, especially since Alberti does leave open the possibility for making this as long as Meliaduse wishes. Two issues then arise. First, the longer AD is, the more dubious it would be that Meliaduse could sight from spear to spear, especially since in his day there were not yet any telescopes or monoculars available for assisting distance vision. Second, is there not an underlying assumption that AE should be substantially smaller than (and thus more readily measurable than) A-Bologna, since if not one could just measure the latter directly? In other words, the more 'well partitioned' triangle AE-Bologna is, the less sense it makes to measure AE as opposed to A-Bologna directly.

We should consider the fact that Alberti's method would be useful in certain circumstances. Say the distance x corresponding to A-Bologna is the distance from your military camp to that of an opponent. You cannot approach their camp, but you can measure out distances within the territory you control that are comparable in length to x, close enough for Alberti's method to work. Or, in another special circumstance, the terrain from A to 'Bologna' could be very treacherous, while the terrain between points A and E could be flat. Had Alberti been thinking in these terms, however, then he should have made some qualifications at the beginning of the problem, as he had in previous problems, when instead he started by instructing: 'Measure any large distance like this'.

Alberti's method is certainly theoretically interesting, but the numbers he supplies simply do not work. It is important to note that we are discussing mathematics and logic well within Alberti's reach. If issues such as the ramifications of specifying a well partitioned triangle were not on his mind, they should have been. In other words, in this case a fault of omission is just as bad as a fault of commission.

Some Alberti scholars, in deference to the Renaissance master, may wish to conclude that these considerations are evidence against the manuscript reconstruction proposed by Furlan and Souffrin. We do not. Although there may never be certainty about the correctness of their reconstruction, either on a philological or a mathematical basis, we certainly have no better explanation as to how to interpret Problem 17 so that it makes sense (at least in theory, if not always in application). Clearly Alberti was trying to invoke similar triangles somehow, and the reconstruction provided by Furlan and Souffrin, in addition to its rigorous approach, is the most logical way to do this.

Problem 18
Constructing an device for measuring lengths of distances along a road (Gal. 10 fols. 14v-15r)

To visualize this device, think of the mechanism by which a gumball machine dispenses a single gumball. A canister contains many gumballs, and through the force of gravity a single gumball fills a hole in a mechanism. When the mechanism is turned (which requires inserting a coin, but this part of the process is not pertinent), then after half a revolution the single gumball falls out of the hole, and after a full revolution the hole is back in position to receive another single gumball. Now imagine the gumballs being made out of a much harder material and much smaller (Alberti's 'pellets'), imagine the appropriately smaller hole and turning mechanism being built into the hub of a wheel (an extension of the axle), and imagine both a canister to hold the balls above the hub and a canister to collect the balls below the hub. This is what is illustrated and described in Problem 18.

The extent to which this device would work is directly related to precision in construction and conditions in practice. How uniformly shaped are the pellets? Is the mechanism precise enough to feed the pellets consistently? How fast is the axle turning? Would ill-timed bumps be problematic? It is hard to imagine that the requisite precision to build a well-functioning device as described existed in Alberti's time.

Vitruvius's hodometer, the last of the non-military devices described in his treatise, is a more sophisticated version of Alberti's description.[116] The hodometer uses gears to convert many revolutions of the actual cart wheel to far fewer revolutions of the counting wheel, at a ratio of 400:1 as described by Vitruvius. The slower speed of the counting wheel makes it much more likely for the stone to be coerced to drop out of a hole. The whole operation seems much more realizable with Vitruvius's construction, although in this solution the issue becomes the construction of the interlocking gears. Of course, Alberti took away the gears in order to make the game more feasible for Meliaduse, but the problem is that the pellet-dropping strategy is much more likely to be workable at slower speeds, even with the technology we have today.[117]

Alberti ends this game (or starts the next – it is hard to tell sometimes) by stating that the contraption of Problem 18 can be used on a ship, as well, by substituting for the wheel a sort of windmill that is turned by the wind rushing past the ship. Vitruvius also describes how to use the hodometer on a ship, but he uses a water wheel instead of a windmill.[118]

---

[116] See [Vitruvius 2009: 300–303 (X, ix)].
[117] See also [Vagnetti 1972: 247].
[118] See [Vitruvius 2009: 301–303 (X, ix, 5–7)].

Problem 19

Constructing an instrument for measuring the speed of a ship using the wind (Gal. 10 fols. 15r-15v)

The greatest significance of this game, at least historically speaking, is that Alberti is widely considered to be the inventor of the first mechanical anemometer, a device for measuring wind speed:

> The first anemometer of any kind that we know about was a swinging-plate instrument, and was described and illustrated by Leon Battista Alberti at some time near 1450. It is in a little work called 'On the pleasures of mathematics'...[119]

Alberti is not concerned solely with measuring wind speed, however, but in measuring ship speed via the wind.

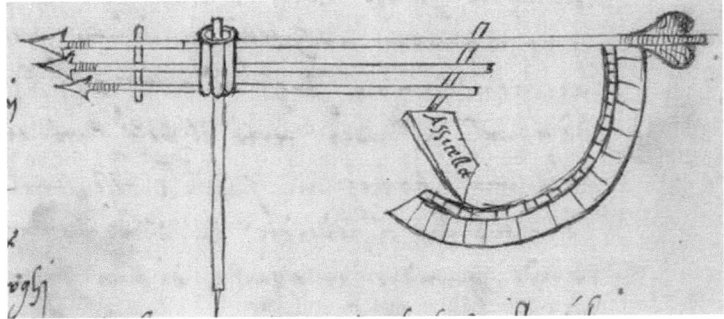

Fig. 45. Gal. 10 illustration for Problem 19

The method comprises a hanging wooden flap built into the rear end of a vane; the flap rises due to wind pressure purportedly correlated to ship speed (fig. 45). One side (the upwind side) of the flap is attached to the vane, and the other side (the downwind side) moves up and down in a swinging motion, the arc of which is mirrored by a curved piece built onto the end of the vane. A scale can be determined on the curved piece, with higher marks correspond to faster wind speed – and, presumably, ship speed, though this may be dubious.

Alberti suggests recording the results of this device under certain known conditions, such as certain configurations of the sails, or given a certain number of rudders in the water, or under certain loads. Building up a repertoire of such information would also enable the user to calibrate the curved piece with marks, at least in theory. Alberti advises: 'and these marks and notations make them so that they are extremely precise and clear'. It seems that this method would work well in practice only if the user is very diligent about keeping records (perhaps in his head, if he has a good memory), and even then, only after quite a bit of experience would the user be able to include Alberti's device as a viable tool in his navigation arsenal.

In fact, Leonardo da Vinci criticizes Alberti's method on just these grounds:

---

[119] [Middleton 1969: 182].

The ancients used various devices to ascertain the distance gone by a ship each hour, among which Vitruvius gives one in his work on Architecture which is just as fallacious as all the others; and this is a mill wheel which touches the waves of the sea at one end and in each complete revolution describes a straight line which represents the circumference of the wheel extended to a straightness. But this invention is of no worth excepting on the smooth and motionless surfaces of lakes. But if the water moves together with the ship at an equal rate, then the wheel remains motionless; and if the motion of the water is more or less rapid than that of the ship, then neither has the wheel the same motion as the ship so that this invention is of but little use. There is another method tried by experiment with a known distance between one island and another; and this is done by a board or under the pressure of wind which strikes on it with more or less swiftness. This is in Battista Alberti.

Battista Alberti's method which is made by experiment on a known distance between one island and another. But such an invention does not succeed excepting on a ship like the one on which the experiment was made, and it must be of the same burden and have the same sails, and the sails in the same places, and the size of the waves must be the same. But my method will serve for any ship, whether with oars or sails; and whether it be small or large, broad or long, or high or low, it always serves.[120]

The footnote at the end of this second paragraph states: 'Leonardo does not reveal the method invented by him'.[121] Leonardo does, however, draw an anemometer similar to Alberti's device in the *Codex Atlanticus*, fol. 675r.

We know from Leonardo's comment on the *equilibra* (see the commentary to Problem 13 above) that he cited the *Ludi* directly, and so it is quite probable that it was his source concerning Alberti's anemometer cum ship speedometer. However, we also know from *De re aedificatoria* of another potential source for Leonardo. In Book V, chapter 12, which is devoted to ships – their types, construction, defence, etc. – Alberti refers to a 'little book'

---

[120] Cf. [Richter 1970: v. 2, 273–274]: *Anno li nostri antichi vsato diuersi ingiegni per vedere che viaggio faccia vn navilio per ciascuna ora, infra li quali Vitruvio ne pone vno nella sua opera d'Architettura, il quale modo è fallace insieme cogli altri; e questo è vna rota da mulino tocca dall'onde marine nelle sue stremità, e mediante le intere sue revolutioni si descrive vna linia retta che rappresenta la linia circunferentiale di tal rota ridotta in rettitudine; Ma questa tale inventione non è valida, se non nelle superfitie piane e immobili de' laghi; Ma se l'acqua si move insieme col navilio con equal moto, allora tal rota resta inmobile, e se l'acqua è di moto più o men velocie che 'l moto del nauilio, ancora tal rota non à moto equale a quel del navilio, in modo che tale inventione è di poca valitudine; Ecco vn altro modo fatto colla sperientia d'uno spatio noto da una isol a vn altra, e questo si fa con un asse o lieua percossa dal uento, che la percuote o più o men velocie, e questo è in Battista Alberti.*
Il modo di Battista Alberti è fatto sopra la sperientia d'uno spatio noto da vna isola a un altra; Ma tale inventione non riesce, se non a un navilio simile a quel dove è fatto tale sperientia, ma bisognia che sia col medesimo carico, e medesima vela, e medesima situation di vela, e medesime grandezze d'onde; ma il mio modo serve a ogni navilio, si di remi come vela, e sia piccolo o grande, o lungo e alto, o basso, sempre serve. Cf. also [Vagnetti 1972: 250].
[121] [Richter 1970: v. 2, 274].

of his called *Navis* (*The Ship*); moreover, in the footnote to this sentence, Joseph Rykwert et al. state, referring to *Navis*: 'Although known in the sixteenth century to Leonardo da Vinci, it was not printed, and no manuscript has come to light'.[122] Mancini's description of *Navis*, however, would seem to indicate that it deals with the same sorts of issues as does *De re aedificatoria* V, vii, which definitely does not mention the contraption of this game.[123] Thus, while we cannot claim to be certain that the *Ludi* was Leonardo's source for Alberti's anemometer/speedometer, it seems very likely.

Problem 20
Measuring volumes of solids by displacement of water à la Archimedes
(Gal. 10 fol. 15v)

This game is Alberti's account of the famous story about Archimedes, the most well-known aspect of which has him exclaiming 'Eureka!' while jumping out of the bathtub, having just thought of the solution to a puzzle that had been occupying his mind. The key to the solution is the fact that the volume of an object is equal to the volume of water it displaces. As Alberti instructs us, this allowed Archimedes to prove a crime concerning 'a certain work of gold' that Prince Hiero had commissioned.

The Vitruvian version, which is virtually identical to the popular version to this day, has Hiero providing a specific quantity of gold to an artist out of which to make a crown. The crown as delivered had the same weight as the quantity of gold, but upon hearing rumours, Hiero comes to suspect that the artist had kept some of the gold and replaced it with the same weight of silver. Having no way to prove this, Hiero turns to Archimedes, who reasoned that since silver is less dense than gold, an equivalent weight of silver would have more volume, and so if the volume of the crown could be shown to be greater than it should be, the crime would be exposed. Archimedes thought of the solution in the bath (a natural place to think about the displacement of water), at which point he hurried home stark naked through the streets shouting *heureka!* ('I have found it!'). He takes an amount of pure gold equal to the amount provided to the artist and an amount of pure silver of the same weight, and then he shows that the silver displaces more water, and therefore has a greater volume, than the gold. The final step is to show that the crown displaces more water than the gold, thereby proving the artist's guilt.[124]

In conveying the story of Archimedes, Alberti does not include quite as much detail in the *Ludi* as Vitruvius does in his treatise, but the basics are the same. He adds some useful information afterwards. Essentially he gives values for the relative densities (or, to be precise, relative specific weights) of wax, pure copper, Cypriot copper, tin, lead, and gold, based on a very particular reference: an amount of wax weighing 1 *oncia*.[125] For example, the same amount of pure copper will be 8 *once*, and so copper is 8 times as dense as wax.

In addition, Alberti rehearses the basic principles of buoyancy, for which he could have referenced Archimedes:

---

[122] [Alberti 1988: 136 (V, xii) and 384, note 44].
[123] See [Mancini 1882: 316–317] for a description of the contents of *Navis*.
[124] See [Vitruvius 2009: 244–247 (IX, introduction, 9–12)]; even the bit about Archimedes being stark naked is in Vitruvius.
[125] Density is mass per unit volume, whereas specific weight is weight per unit volume.

So then any item being of equal measure with the water and in itself weighing less, this will be lifted up and kept afloat by as much as its weight is less, and it will stay likewise immersed in the water by as much as the quantity of water will be of a weight equal to it.

Here 'equal measure' means equal volume, so that the different weights are directly comparable. Note that the key implicit notion is the density of the object relative to water: the density of the object must be less than the density of water, in order for the object to float.[126] By the same token, objects denser than water will sink, the denser the faster, as Alberti states in so many words. He then mentions having used this principle while working with some architects to determine how much a certain column weighed.

The principle of buoyancy discovered by Archimedes, which Alberti seems to have known fairly well, leads to a refinement of the golden crown story. The drawback to the method described by Vitruvius is that the small difference in the amount of water displaced by the gold and the crown may be hard to measure. The refinement is to use a scale (such as Alberti's *equilibra*; see Problem 13) to first balance the equal weights of gold and the crown. Then dip the whole balance into the water, and the side with the gold will tip (assuming the crown really is less dense).

This link between Problems 13 and 20 is not without precedent in the literature. Thomas Settle, in an article on Ostilio Ricci's role as a bridge between Alberti and Galileo, in particular via Ricci's probable use of the *Ludi* as a text for teaching, states:

> Two items in the *Ludi* especially pertain to one of Galileo's very early works. The first has to do with the measurement of small weights. Alberti shows how to use the same arrow mentioned previously, along with a bit of string, to make a simple but convenient and accurate balance. The techniques, I think, are related to the ones Galileo describes in his first paper, *La Bilancetta*. Then in a short section at the very end of the *Ludi*, Alberti repeats the story of Archemedes [sic] and Hiero's Crown. He also describes density measurements in general and compares the known relative weights of some materials. Now we know that the story of Hiero's crown was common currency in the 16th century, but it would not be too far fetched to suggest that when Galileo refers to it, again in *La Bilancetta*, he actually has this particular text in mind.[127]

Souffrin makes this same link between Alberti's *Ludi* and Galileo's first published scientific work, *La bilancetta* (1586), in which Galileo contests the account of Vitruvius and describes his own solution of the problem using a variant of the hydrostatic scale of his invention. Because of such issues surrounding this spectacle, Souffrin states that Problem 20 of the *Ludi* is the one that has generated the greatest interest among the historians of physics.[128]

---

[126] The density of an object relative to water is called its specific gravity (not to be confused with specific weight), which must be less than 1 for the object to float.

[127] [Settle 1971: 123]; for more about Ricci, see Angela Pintore's notes on the transcription, p. 71 in this present volume.

[128] See [Alberti 2002: 109ff] for an extended discussion, including a translation (into French) of Galileo's *La bilancetta*. More recently, studies of the golden crown problem have run the gamut from *Scientific American* to Wikipedia.

Concluding remarks

Ending with a well-known tale could be interpreted as Alberti trying to finish up and get the job done to satisfy Meliaduse, as Vagnetti suggests in the concluding remarks of his commentary.[129] But perhaps Alberti had intended to end with a great, historically relevant spectacle. To bolster this argument, consider that Alberti was thinking along the way about the order of the *Ludi*; indeed, towards the end of Problem 3, Alberti foreshadows future games:

> If you want to measure the height of a tower that you can't get close to, but you can see its base and its top well, it is best to find a way to know how much the space between you and the foot of the tower is, because if you know this space for sure, then from the measures given above you will be well able to understand how high it is. There is a way to know this distance that I will set forth here below, suitable for measuring any distance, especially when it isn't very far away. To measure those quite far way, I will give you a singular way.

As we state at the end of the Problem 3 commentary, it seems most likely that Alberti is foreshadowing Problems 6 and 17, though it is possible he is referring to Problems 17 and 18. In either case, Problem 17 is involved, which is towards the end of the *Ludi*. This shows that Alberti was consciously choosing some sort of order for the games; perhaps he always planned to end with the tale about Archimedes.

The first seven problems concern the ways in which similar triangles can be used to measure heights, widths, and depths. Problems 1 and 6 deal with two pairs of similar triangles, i.e., similar within the pairs but non-similar between the pairs. Problem 6 requires the most sophisticated mathematics to find the desired length, given the known quantities, so it is not surprising that this is the one of the first seven that has suffered the most corruption by copyists. Beyond these first seven, a few others also use similar triangles, namely Problems 13 (at least implicitly), 16, and 17, which means that half of the games involve them. Lines of sight, what Alberti calls visual rays in *On Painting* (see the commentary to *Elements of Painting*), make up one or more of the sides of the triangles used in these problems. The concomitant workhorse, in a mathematical sense, corresponds to Euclid's *Elements* VI, Prop. 4, whereby similar triangles lead to proportions involving their sides' lengths.

Basic geometry also plays a vital role in some other problems, certainly Problems 12 and 15, and to a lesser extent Problems 10 and 13. Problem 13 is a bit of a hybrid, and we can see why Souffrin divides it into three separate games in his list, at least based on the amount of content in our commentary on this set of exercises. The rest of the problems have more of a science and engineering approach, dealing with mechanics, astronomy, fluid motion, etc. Recalling that *ludo* can mean spectacle, we would maintain that this grouping of problems, in particular though not exclusively, show evidence of Alberti trying to impress Meliaduse.

To the extent that it was possible, Alberti chose materials that would be readily available to Meliaduse, even those that he might have with him in the field, so that he could apply some of these methods spontaneously if needed. The spear (*dardo*, which

---

[129] See [Vagnetti 1972: 256].

could also be translated as 'dart' or 'arrow') is a common companion in these games, showing up in half the problems. It is used as a sort of surveyor's pole to sight points against in Problems 1 through 6, as a pole to cast a shadow in Problem 10, as the arm in a hanging balance (the *equilibra*) in Problems 13 and 15, and again as a sort of surveyor's pole in 17 (though in this case no points on the spears are marked). String or cord is used to make a makeshift compass in Problem 10, to create a 3-4-5 right triangle in Problem 12, and to construct the *equilibra* used in Problems 13 and 15. Wax, used to mark points on a spear, is another example of an ordinary material incorporated into the *Ludi*. There are some exceptional games, which use fairly elaborate materials and/or mechanisms, most notably Problems 14 and 18. But in general, the materials Alberti chose for the *Ludi* can be seen as analogous to the fact that he took care to provide methods that involve fairly basic mathematics.

Unfortunately though, as we have seen, Alberti does not always do the best job of seeing a problem all the way through to an understandable conclusion. In many instances it seems that Alberti is torn between wanting to keep things simple enough for Meliaduse and wanting to show off the interesting mathematics. In some cases, he writes that a given topic is too complicated, not fitting for the approach of the mathematical games, but then he proceeds to describe the mathematics in at least some detail, in case Meliaduse is interested after all. In other cases he ends abruptly, without filling in enough details for Meliaduse to be able to use the method in the general case.

As we state in the introductory remarks to this commentary, measuring quantities such as heights, lengths, widths, weights and depths (which are continuous as opposed to discrete) inherently involves approximation, i.e., rounding to the nearest chosen unit. Moreover, many of the games are either approximate in and of themselves or exact in theory but clearer prone to error in practice. Presumably, however, it would not be critical for Meliaduse to know, for example, that a river is 29 *braccia*, 7 *once* and 9 *punti* wide – it would probably serve his purposes to know it was on the order of 30 *braccia*.

Still we must not assume that Alberti was unconcerned with precision. Indeed, consider Alberti's advice to Meliaduse in Problem 12: 'And note that the square needs to be really large if you want to have a good degree of certainty. The large square errs less'. This language is echoed in *De re aedificatoria*:

> Therefore, before you start any excavation, it is advisable to mark out all the corners and sides of the area, to the correct size and in the right place several times, with great care. In setting out these angles use an extremely large set-square, and not a small one, to make the direction of the lines more accurate. The ancients used a triangular set-square consisting of three straight rules, one three cubits in length, another four, and the other five.[130]

There are various instances in *De re aedificatoria* where Alberti discusses precision and provides ways of ensuring it.[131]

---

[130] [Alberti 1988: 62 (III, i)].

[131] See, for another example, [Alberti 1988: 314 (IX, ix)]: 'In short, everything should be measured, bonded, and composed by lines and angles, connected, linked, and combined—and that not casually, but according to exact and explicit method.... I need not repeat the errors of craftsmanship; but make sure that the workmen make proper use of the plumb line, string, rule, and set square'.

Vagnetti and Souffrin both comment on the lack of precision in the *Ludi*. Vagnetti repeatedly does so, noting obviously that there was less precision possible in Alberti's time than there is now, seemingly as a way to excuse or defend him. Souffrin's analysis seems to us to be more to the point. After discussing Problems 14 and 17, which he states are probably Albertian in origin and clearly have issues with precision in practical realization, Souffrin states:

> What the *Ludi* would deal with would be showing theoretically the singular capacity of mathematics to allow *in principle* to resolve without *another* tool, without *specific instrumentation*, the problems of measurement that the man of that time encounters in practical life.[132]

Perhaps our favourite perspective on this matter, however, is that expressed by Settle:

> What I am saying, in part, is that the *Ludi* is not only a manual for measurements, it is also an invitation to measure, an advertisement, if you will, for the pleasures and possibilities open to those respecting mathematical precision in the physical world.[133]

In our opinion, this captures the essence of Alberti's games, his amusing spectacles, his pleasurable exercises, his delightful toys. The point is that Alberti was trying to showcase mathematics, and he most certainly succeeded.

---

[132] Cf. [Souffrin 1998: 91]: *Ce dont il s'agirait dans les* Ludi, *ce serait de montrer théoriquement la capacité singulière des mathématiques de permettre* en principe *de résoudre sans* autre *outil, sans* instrumentation spécifique, *les problèmes de mesure que rencontre dans la vie pratique l'homme de ce temps* (trans. María Celeste Delgado-Librero).
[133] [Settle 1971: 124].

Elements of Painting

K. Williams et al. (eds.), *The Mathematical Works of Leon Battista Alberti*,
DOI 10.1007/978-3-0346-0474-1_3, © Springer Basel AG 2010

ELEMENTS OF PAINTING

Have you ever seen a blind man teach the way to someone who sees? Here with these briefest of notes, which we call Elements, you will see that someone who perhaps does not himself know how to draw can show the true and sure reasoning and the way to become a perfect draughtsman, as long as you don't shy away from learning that which you judge to be impossible. First try to see if you can do it, and then judge our erudition and your acumen as you will. Trying, you'll believe me, and believing me you'll take delight in knowing them all. Do like this, and love me.[1]

---

[1] Translator's note. This short introduction precedes the Italian version of *Elementi di pittura*. Compare it to the longer introduction, the letter to Theodore of Gaza, in the Latin version, *Elementi pitturae*.

To Theodore of Gaza

Do you really think that it can possibly happen anywhere, Theodore, that someone who is himself absolutely incapable of seeing, could show the road on which those who can see, but do not know the way, should set off.[2] We should realise that with these Elements of mine – for that is what I call these brief suggestions – those who understand them, though they may be rough and inexperienced in other respects, would at least have the means by which they could instruct those who are enthusiastic and eager to learn about painting with little effort, and could readily turn them into the type of people that the most erudite would be accustomed to praise, so long as they do not fail to appreciate that, before the subject has been understood, it would probably seem hardly credible. So I think they should be alerted to the fact that, before they concentrate on what it is we have in mind to do, they should look immediately at whether the matter in hand results from common practice, then they should judge us and themselves and decide whether they want to do it. For when they understand from what unimpeachable sources these rules have been derived, they can believe whatever they like about the degree to which I have expended so much labour in vain in dealing with these extremely abstruse and recondite matters, and they will be so far from regretting the labour that they should be congratulated even more that they themselves have been made the equals of outstanding painters by using this advice and their own diligence, and how much better they could understand the subject by practicing it than by any possible explanation in words by me. For it will be obvious what advantages these writings would bring, what facility they would produce, how much they would guide the hand, the eyes and the mind to understand and retain absolutely trustworthy and easily-applied principles of painting. This I assert lest anyone who has ignored these rules ends up with a place among the really ordinary painters. So it is worthwhile issuing this warning to those to whom these things might seem all too remarkable. But it was not necessary to warn you, a man of outstanding competence in all branches of learning, of such matters. But when you had said so often that my three books On Painting had pleased you and had asked that I should also put into Latin these Elements, which I had previously published in the Etruscan language for the sake of my fellow citizens, and that I should send them to you to study, I wished to satisfy fully your expectations and our friendship to the best of my abilities. So I translated them into Latin and even dedicated them to your name, which should be something fortunate and happy and a monument and everlasting pledge of our friendship. Take then the Elements and Leon Battista, your most obedient servant for as long as you wish, and love me as much as I myself would wish; for I want you to love me – as much as you do – deeply.

---

[2] The dedication to Theodore of Gaza was translated from the Latin by Richard Schofield.

ELEMENTS

A. In order to be brief in writing and clear it occurs to me to set forth these definitions here:[3]

1. The point is said to be that which cannot be divided at all into any parts.
2. A line is said to be almost <like>[4] a point stretched out in length. Thus the longitude <length> of a line can be divided, <but> not its latitude <width>.
3. A surface is said to be produced[5] almost as though extending the line width-wise, and thus its length {and width}[6] can be divided, <but> not its depth.
4. A body is said to be whatever can be divided by length, by width, and also by depth.

B. These <statements above> were stated by the ancients.[7]

1. We call body that which is covered by surfaces, on which our vision comes to rest.[8]
2. We call surface that extreme skin of the body, which is encircled by its edge.
3. We call edge the whole circuit, almost <like> borders, where the surface ends, which place I call *discrimen,* a term taken from the Latins.[9]
4. A *discrimen* proper is the line drawn from the forehead that divides the hair so that some falls to one side of it and some to the other. Thus using this simile, in our case *discrimen* will be that length in the middle between two surfaces that divides one from the other, and this length terminates in two points.[10]
5. I call point that extremity where several lengths and *discrimens* share a common termination.[11]

---

[3] Latin text adds: *has praeposuisse diffinitiones iuvet sumptas ex mathematicis* (it should help to have presented first these definitions derived from mathematicians). This and all the translations from Latin that follow are by Richard Schofield.

[4] Integrations to the text for the sake of clarity are added angle brackets < >; integrations included in the 1973 Grayson edition are added in {} brackets.

[5] The verb used here is 'addurre', which in this first instance we translate as 'produce', while in the three others instances it is used (B.4; D.5; D.7), we translate it as 'draw'.

[6] The Grayson edition correctly integrates this term, which is included in the Latin text: *atque item latitudo* (and width); certainly this insertion clarifies Alberti's meaning.

[7] Latin text adds: *Nos ista subiungemus* (We should add these).

[8] Lat text differs: *corpus appello id quod opertum superficie sub aspectu et lumine possit perspici* (I call a body that which is covered with a surface and can be perceived by the power of sight and vision).

[9] Latin text omits the final phrase.

[10] Latin text abbreviates the paragraph: *Discrimen, quod ex capillorum similitudine duximus, ea est finitio superficierum qua altera ab altera secernatur, ducta a punctis conterminalibus* (Discrimen, a term which we have derived from a simile with hair, is that border between surfaces by which one is separated from the other, drawn between end-points).

[11] Latin text adds: *qualis in adamante est cuspis* (like the point of a diamond).

C. These considerations <above> are ours. Concerning our painters:[12]

1.   In painting we call point that small inscription than which nothing can be smaller.

2.   We call line that very fine inscription that goes from one point to another, by which an area is circumscribed by its edge.[13]

3.   We call edge the entire outlined description made of lines most fine, by which one area is divided from another.[14]

4.   We call area that space the edge of which we circumscribe in a certain similitude to the surface viewed.[15]

D. Add to these that:

1.   A concentric[16] area we call that which, with its angles and lines, is enclosed by the whole of the edge, and thus corresponds to that surface seen from a certain view point, such that it does not therefore appear in any of its parts to be smaller than it actually is, in comparison to the others.

2.   A comminuted[17] area we call that which is similar to the surface viewed, when it is placed in such a way that it will appear smaller in some of its parts.

---

[12] In the Latin text the last phrase differs slightly: *nunc quae ad opus pingendi faciant* (now for what they should do with respect to the practice of painting).

[13] In the Latin text the last phrase differs: *quibus pictor linbum areis circumscribat* (by which a painter should circumscribe an edge on areas).

[14] This paragraph has no corresponding Latin text.

[15] Latin text C.3 differs slightly: *Aream appello id spatium in pictura, quo visae superficei amplitudinem certis lineis et angulis imitemur* (I call area that space in a painting by which we represent the extent of a surface observed with particular lines and angles).

[16] The entire Latin text is worth including: *Concentrica in corpore superficies est, quae non mutato intervallo sub aspectu ita extat, ut maior nullo modo alio videri possit. Concentrica igitur erit area in pictura, quae istam repraesentet* (A concentric surface on a body is that which, without any change in distance, appears to the gaze in such a way that that it cannot seem greater in any other way. Thus an area in a painting which represents this will be concentric). Alberti's use of the term 'concentric' does not correspond directly to our standard modern usage (having a common centre, having a common axis); for further discussion see the commentary.

[17] In the Latin text Alberti adds some refinements to this definition: *Comminutam dicemus superficiem hanc, quae sub aspectu ita sit posita, ut aliqua seu linearum seu angulorum inter se comparatione minor parte aliqua sui esse videatur quam re ipsa sit. Comminuta itidem erit area in pictura, quae istanc exprimat* (We call a comminuted surface that which is placed in relation to the gaze in such a way that some of either its lines or its angles, in comparison with each other, appear to be smaller in some part than the thing itself. In the same way an area in a painting which expresses that will be comminuted). Alberti's use of the term 'comminuted' does not correspond to our standard modern usage (fractured or disintegrated); for further discussion see the commentary.

3.   A proportional area will be when each of its lines will be either longer or shorter in some given part of it than those of the viewed surface appear to be.[18]

And we further consider:

<...>[19]

4.   A commensurate point will be that which will have coequal distances from the other points, whether they are of concentric or comminuted <areas>.[20]

5.   A straight line is that stroke drawn from one point to another by the shortest way, and which does not curve in any of the spaces.[21]

6.   Thus those lines or areas or surfaces that do not have any curved lines at all are called rectilinear; and so curvilinear are those in which there are curved lines.[22]

7.   We call a curved line that which, drawn from one point to another, describes a certain part of some circle.[23]

8.   Here by circle we want to indicate an edge that is continuous and constructed of some number of curved lines joined to one another at their extreme ends, such that they neither intersect nor create any angles; and the edge[24] should be constructed such that none of its parts are closer to a centrally placed point than the others.

9.   An angle is where two joined lines do not constitute one line, but intersect each other; and of these some are not right and some are right. A right angle is one of the four angles constituted by two crossing lines that intersect each other which will be neither larger nor smaller than any of the other three. A non-right angle

---

[18] The Latin text is more extensive: *Proportionalis erit area, seu comminuta illa quidem seu concentrica sit, quae lineis aut maioribus aut minoribus conscribetur quam ut aequent certo sub aspectu positam superficiem, in caeteris omni dimensionum comparatione partes partibus correspondebunt* (An area will be proportional, or at least somewhat comminuted or concentric when it is circumscribed by longer or shorter lines in such a way that they make level the surface laid down from a certain point of view, and in the rest, the parts will correspond to every combination of the dimensions).

[19] Latin text D.4 omitted in Italian version: *Compar erit area quae tanta sit quanta esse ampla sub certo aspectu posita superficies videatur* (An area will be equal when it is as large as a surface laid down appears to be from a certain point of view).

[20] Corresponds to Latin text D.5: *Punctum commensuratum in pictura erit cum a caeteris picturae punctis certa intervalli ratione distabit* (A commensurate point in a picture will be one with distances in a certain relationship to other points in the picture). See the commentary for an extensive discussion of this definition.

[21] Corresponds to Latin text D.6; in the Latin text the final phrase differs: *qua nulla possit dari brevior* (which cannot be made any shorter).

[22] Corresponds to Latin text D.7; the Latin text adds: *et mixtae quae ex his mixtis* (and mixed areas are those which include both).

[23] Corresponds to Latin text D.8, which adds here: *nam cocleas quidem et columnarum conicarumque sectionum lineas pictor non habet qui imitetur nisi flexarum rationibus et adminiculis* (for the painter does not have the means to represent spirals in particular and the lines of columns and conical sections unless he uses the techniques and methods of curved lines).

[24] Corresponds to Latin text D.9; Latin text differs slightly: *Quod si in areae medio adsit punctum, id ab universis limbi partibus aequo semper intervallo distabit* (So that if there is a point in the middle of the area, it would always be equidistant from all points of the border (or 'edge')).

will be that which is larger or smaller than the right. Thus rectangles or non-rectangles are so called according to whether their angles are right or non-right. And similarly a triangle and a quadrangle and so forth take their names from the number of their angles.[25]

E. The definitions given up to this point are sufficient.[26]

In what follows we set forth in order the principles that are necessary to understand well the true reckoning of drawing.

1. Drawing a straight line from one point to another point.
2. Dividing the space between two points into certain parts with certain points.
3. Extending a straight line so that it is longer by a certain part of its quantity.
4. Drawing a straight line that is equidistant from another straight line.
5. Forming a right angle at a given point.[27]
6. Drawing a right-angle triangle equal to another given triangle.[28]
7. Drawing a non-right-angle triangle equal to any given triangle.[29]
8. Drawing any kind of concentric rectilinear <area>.[30]
9. Inscribing a commensurate point in a rectilinear and concentric area.[31]
10. Placing in one drawn rectilinear and concentric area another smaller concentric and rectilinear <area>.[32]
11. Drawing a commensurate point outside of a rectilinear concentric <area>.[33]
12. Enclosing a drawn rectilinear area with a larger area that is rectilinear and

---

[25] Corresponds to Latin text D.10.

[26] Latin text continues: *Sequitur ut rem aggrediamur. Ex his quae sequentur, omnis ratio et via perscribendi componendique lineas et angulos et superficies explicabitur notaque reddetur adeo ut nihil in rerum natura sit, quod ipsum oculis possit perspici, quin id hinc instructus perfacile possit lineis perfinire atque exprimere* (Now we should get to the point. In what follows, every principle and method of circumscribing and constructing lines, angles and surfaces will be explained and made clear so that there will be nothing in the natural world that the eyes can see that a person instructed in these things cannot readily define and illustrate with lines).

[27] Latin text adds three new paragraphs at this point:

E.6 *ex data linea plures inter se compares partes abscindere unis earum capitibus conterminantes punctis ubi libuerit in ea signatis* (on a given line divide up a number of sections of equal length coterminous with the end of each marked with points on it where you want);

E.7 *datis duabus lineis in diversum protractis ab utrisque partes inter se compares abscindere* (on two given lines drawn out in different directions mark off equal sections from each);

E.8 *ex lineis pluribus ab uno dato puncto in quamvis partem protractis abscindere partes mutuo inter coaequales* (on a number of lines drawn in any direction from a given point mark off equal sections):

[28] Corresponds to Latin text E.9: *Perscribere triangulum rectangulum dato triangulo rectangulo comparem* (Draw a right-angle triangle equal to a given right-angle triangle)

[29] Corresponds to Latin text E.10: *Triangolo cuivis dato alterum comparem describere* (draw a second triangle equal to any given triangle)

[30] Corresponds to Latin text E.11: *Qualemcumque concentricam datam superficiem rectilineam compari area exprimere* (Draw any given concentric rectilinear surface with an equal area)

[31] Corresponds to Latin text E.12.

[32] Corresponds to Latin text E.13.

[33] Corresponds to Latin text E.14.

concentric.

13. Given a rectilinear and concentric surface, drawing an area that is similar but proportionally larger.[34]

14. And drawing a rectilinear concentric area but proportionally smaller.[35]

15. Drawing a commensurate point in a {rectilinear} area that is proportionally {greater}.[36]

16. In a given <area> that is concentric, angular and proportionally larger, inscribing another proportionally larger angular <area>.[37]

17. And likewise, in an angular <area> that is concentric and proportionally smaller, drawing another <area> that is concentric and proportionally smaller.[38]

18. Enclosing a concentric, angular and proportionally larger <area> in one that is concentric and proportionally larger.[39]

19. And enclosing a proportionally smaller, concentric <area> in one that is concentric and proportionally smaller.[40]

F. This is sufficient regarding rectilinear concentric <areas>.

Regarding curvilinear concentric <areas>:[41]

1. Given a curved line, extracting from it another curved line.

2. Given a curved lined, drawing another curved line equal and concentric.

3. Producing a curved line that is somewhat smaller but proportional to a given curved line.[42]

4. Producing a curved line that is proportional to but somewhat larger than that

---

[34] Corresponds to Latin text E.16.

[35] Although only slightly different from paragraph E.13 (with the subsitution for 'smaller' instead of 'larger', this paragraph has no corresponding Latin text.

[36] Corresponds to Latin text E.17: *Intra rectilineam proportionalem aream maiorem punctum commensuratum adnotare* (Mark a commensurate point in a rectilinear area that is proportionately larger). Grayson's integrations of {rectilinear} and {greater} are taken from the Latin text.

[37] The Latin text inserts 4 paragraphs here:

E.19: *Extra proportionalem maiorem rectilineam et concentricam aream punctum commensuratum adnotare* (Mark a commensurate point outside a proportionally larger rectilinear concentric area);

E.20: *Area proportionali maiore concentrica et rectilinea alteram istiusmodi aream circumcludere* (In a proportionately larger concentric and rectilinear area describe another area of that type);

E.21: *Proportionalem aream quota sui parte minorem exscribere* (Draw a proportional area smaller in proportion to the whole);

E.22: *Minorem intra concentricam proportionalem aream rectilineam punctum commensuratum adnotare* (mark a commensurate point inside a smaller concentric proportional rectilinear area).

[38] Corresponds to Latin text E.23.

[39] This has no corresponding paragraph in the Latin text.

[40] Corresponds to Latin text E.25.

[41] Latin text adds: *De ratione subducendi, scribendi, similes faciendi lineas et superficies angulares* (On the method for extracting, drawing out and making similar lines and angular surfaces).

[42] Corresponds to Latin text F.4.

given line.[43]

5. Drawing an area equal to a certain part of a surface, which is cut only by a certain curved line.[44]

6. And drawing an area that is equal to a part of a circular surface cut by a number of curved lines.[45]

7. Drawing a curvilinear area but with a certain larger proportion.[46]

8. Again, drawing a curvi{linear} area that is concentric and proportional but smaller.[47]

9. Placing a commensurate point in a concentric curvilinear <area>.

10. And, in a concentric curvilinear <area>, drawing another concentric curvilinear <area>.[48]

11. Placing a commensurate point outside a concentric curvilinear <area>.

12. Closing a concentric curvilinear <area> with another larger concentric curvilinear <area>.[49]

13. Marking a commensurate point inside a proportionally larger curvilinear area.[50]

14. Placing a curvilinear <area> in any proportionally larger concentric area.[51]

15. Marking a commensurate point outside a proportionally larger curvilinear area.[52]

16. And when there is a proportionally smaller <area> likewise placing inside it the commensurate point.[53]

17. Again, likewise placing the commensurate point outside of a proportionally smaller area.[54]

G. And this is sufficient regarding concentric curvilinear <areas>.[55]

1. Drawing a semicircle above a given line.

2. Finding in any given circle its centre and diameter.

3. Drawing a concentric circle.

4. And drawing a proportionally larger concentric circle.

5. Drawing a circle that is still concentric but with a certain smaller proportion.

---

[43] Corresponds to Latin text F.3.

[44] This paragraph has no corresponding text in Latin.

[45] Corresponds to Latin text F.5.

[46] Corresponds to Latin text F.6.

[47] Corresponds to Latin text F.7.

[48] The previous paragraph F.9 and this paragraph F.10 combined correspond to Latin text F.8.

[49] The previous paragraph F.11 and this paragraph F.12 combined correspond to Latin text F.9.

[50] Corresponds to Latin text F.10, which adds: *ex quo et area istiusmodi inscribatur* (from which an area of the same type can be drawn).

[51] This paragraph has no corresponding text in Latin.

[52] Corresponds to Latin text F.11, which adds: *ex quo altera istiusmodi area circumcludatur* (from which (i.e., the point) another area of the same type can be drawn).

[53] Corresponds to Latin text F. 12.

[54] Corresponds to Latin text F. 13.

[55] Latin text continues: *De ratione scribendi semicirculos et circulos compares, concentricos atque commensuratos* (On how to draw semicircles and equal circles, both concentric and commensurate).

6. Drawing a commensurate point in a concentric circle.
7. And drawing in a concentric circle another smaller concentric circle.
8. Placing a commensurate point outside the concentric circle.[56]
9. Encircling a drawn concentric circle with another circle that is concentric and larger.[57]
10. Placing a commensurate point in a circle that is proportionally larger and still concentric.[58]
11. And placing a commensurate point outside of a proportionally larger circle.[59]
12. Placing a commensurate point in a proportionally smaller circle.[60]
13. And, likewise, placing a commensurate point outside a smaller <circle>.[61]
14. Drawing inside and drawing outside, as you wish, a proportional circular <area> either larger or smaller.[62]

I. And this is sufficient regarding concentric circles.

Up to this point we have said everything that we reckon necessary for all concentric surfaces and areas, which you have seen to be of three kinds: either angular rectilinear, or angular curvilinear, or circular. In what follows we set forth what we foresee to be necessary for comminuted areas and surfaces. Regarding comminuted rectilinear <areas>:

1. Drawing a rectangular <area> similar to a comminuted surface in a rectangular concentric area.
2. In any drawn angular concentric area, drawing another smaller area that is similar to a rectilinear comminuted surface.
3. Circumscribing a concentric <area> with a comminuted angular area.
4. Inside a comminuted angular area, drawing a commensurate point.
5. Drawing in a comminuted angular <area> another comminuted angular <area>.
6. Drawing a commensurate point outside the comminuted angular <area>.
7. Circumscribing a comminuted angular <area> with other angular <area> that is also comminuted.
8. Drawing a proportionally larger area that is comminuted and angular.[63]
9. And drawing an angular area that is proportionally smaller and comminuted.
10. In any of these proportional <areas>, whether larger or smaller, placing the

---

[56] Latin text G.8 continues: *ex quo hunc maiori concludas circulo concentrico* (from which (ie. point) you can enclose this with a larger concentric circle)
[57] This paragraph has no corresponding text in Latin.
[58] This paragraph has no corresponding text in Latin.
[59] This paragraph has no corresponding text in Latin.
[60] Corresponds to Latin text G.9, which continues: *ex quo minorem istiusmodi quoque aream inscribas* (from which you can draw a smaller area of the same type).
[61] Corresponds to Latin text G.10, which continues: *ex quo et circulo itidem proportione maiori hunc circumcludas* (from which you can circumscribe this with a circle of greater size in the same way)
[62] Corresponds to Latin text G.11, which continues: *ex quo et inscribere et circumcludere areas possis* (from which you could draw and circumscribe areas)
[63] Latin text differs: *scribere aream persimilem superficiei comminutae angulari rectilineae, sed quota sui parte maiorem* (draw a similar area to a comminuted, rectangular surface but proportionately larger).

commensurate point.[64]

11.	And likewise, in any of these proportional <areas> inscribing one that is also proportional and comminuted.

12.	And outside any proportional comminuted <area> drawing a commensurate point, from which every similar proportional and comminuted surface can be enclosed.

L. And this is sufficient regarding comminuted and rectilinear <areas>.

Regarding comminuted curvilinear {and}[65] circular <areas>.

The comminuted curvilinear <areas> will have that same order and manner of being taught as rectilinear <areas>, since the curvilinear are deduced from the rectilinear. Thus for the sake of brevity I do not write it down.[66]

1.	Drawing a comminuted circular area in a given rectangular <area>.

2.	Drawing a comminuted circular area in a given angular <area>.[67]

3.	Circumscribing the concentric angular area with a comminuted circular <area>.

4.	And circumscribing a comminuted angular <area> with a<nother> comminuted angular <area>.

5.	Inscribing a comminuted circular area in any one that is an angular area.

6.	Placing the commensurate point in the comminuted circular area.

7.	Placing the commensurate point outside the circular and comminuted area.

8.	Drawing a comminuted circular area in another area that is also circular and comminuted.

9.	Circumscribing with a comminuted circular <area> an area regardless of whether it be angular or circular.[68]

10.	And drawing a circular area that is comminuted and proportionally larger.

11.	Inside this proportionally larger <area> placing the commensurate point, and there drawing a whole circular <area> that is comminuted and proportionally larger.

12.	And drawing a circular <area> that is proportionally smaller and comminuted, and in this placing the commensurate point, and there drawing a whole <area> that is proportionally smaller and comminuted positioned inside it.

13.	Drawing any area in any other area.

---

[64] Latin text differs: *in data proportionali et comminuta angulari rectilinea maiore punctum adnotare commensuratum; et in data proportionali minore comminuta angulari rectilinea punctum commensuratum adnotare* (mark a commensurate point in a given larger, angular and rectangular area; and mark a point in a given smaller, comminuted angular, rectangular area

[65] We agree with Grayson's integration.

[66] Instead of last sentence, Latin says: *aliqua tamen referentur, quae angularibus flexilineis conferant* (other information should be mentioned that helps with angular curved lines).

[67] Latin text differs: *circularem aream comminutam intra rectangulam concentricam scribere* (draw a circular, comminuted area in a concentric rectangle).

[68] Latin says differs: *circumcludere aream circularem comminutam altera circulari* (circumscribe a circular, comminuted area with another circular one).

14.   In proximity to any area, drawing another area similar to whatever kind of surface viewed.[69]

M. And this is sufficient regarding comminuted circular areas.

I beg of you, o scholars of my things, strive to understand these Elements, which perhaps by <mere> reading do not reveal themselves to you. I promise you that you will be pleased to have worked on it, and you will see that this doctrine is no less useful than it is amusing.[70] Love your Leone Battista, who hopes in still better things to be useful to your studies. I beg whoever transcribes for me to reread it for me and with diligence make emendations.[71]

---

[69] Latin text L.13 is more ample than paragraphs L.13 and L.14 in Italian: *Qualescumque dederis superficies, seu concentricas, seu comminutas, alteras alteris inclusas, seu exclusas, seu coniunctas, seu disiunctas persimiles eis areas aut velis compares aut velis proportionales exscribere, et qua id ratione viaque effeceris, monstrare* (Demonstrate how to draw whatever surfaces you will have proposed, whether concentric, comminuted, with others inside or outside them, connected or not connected to them and areas similar to them, or , if you wish, equal or proportional, and on what principles and methods you will achieve this.

[70] The word is *iocunda*, modern *gioconda*, jolly or amusing, in the same sense as 'Ludi matematici'.

[71] Latin text differs: *Quae circa Elementa dicenda videbantur transegimus, et sunt quidem ea, ut vidisti, eiusmodi ut a notissimo perfacilique principio ad ultimam usque atque penitus reconditissimam istius artificii rationem et cognitionem adducant. Sed agendo altera ex alteris percipiantur oportet. Quare obsecro qui nostris inventis delectentur, omni studio et diligentia instent ac prosequantur quoad totam hanc eruditionem prehenderint. Ex ipsa enim re perspicient iucundam esse discendi viam non minus quam utilem. Peto etiam ab his qui exscripserint opusculum hoc, diligenter emendent et numeros admonitionibus adiungere non negligant* (We have gone through the things which it seemed important to say about the Elements, and they are in fact things which, as you have seen, are such that lead from a perfectly well-known and simple principle to the ultimate and most deeply recondite theory and understanding of this art. But it is inevitable that by putting them into practice new principles would be understood from others. Which is why I beg those who take delight in my discoveries to press them with the greatest attention and diligence and pursue them so that they understand this whole subject completely. For as a result of this they will realise that the road of erudition is as delightful as it is useful. So I ask those who will be transcribing this work that they should correct it carefully and not forget to add numbers to each of the instructions).

Commentary on *Elements of Painting*

Stephen R. Wassell

Alberti first wrote his *Elements of Painting* in Italian as *Elementi di pittura*, probably between 1432 and 1435, and sometime thereafter he wrote a Latin version entitled *Elementa picturae*, probably between 1435 and 1446.[1] The dedications of the two versions differ substantially, with that of the Latin version being over four times as long, but the subsequent content is largely the same. *Elements of Painting* is a short treatise, comprising about 2000 words, arranged as a succession of lists, with many items consisting of fewer than twenty words. The first four lists provide the definitions of terms that Alberti uses, and the remaining lists offer exercises concerning the various drawing skills that the practitioner must master as a prerequisite to becoming a good painter. It is important to realize that painting, for Alberti, inherently assumed realism and hence benefitted from the then-nascent use of techniques that we would now call perspective.[2]

While it can be problematic to attempt to infer the true inspirations and intentions that motivated Alberti's writings, it is virtually certain that he developed much of the content of *Elements of Painting* according to Euclid's *Elements*, and that Alberti chose its title and mathematical format in reference, and possibly deference, to the classical Greek text. It is instructive, in this regard, that in the dedication Alberti himself refers to his work as 'these briefest of notes, which we call Elements'. Moreover, it is known that he was a careful reader of Euclid's *Elements*, his own copy of which he extensively annotated in the margins.[3]

---

[1] Exact dates are not known, though it seems certain that the overall time frame of 1432 to 1446 is correct; see note 4 below for details.

[2] The Latin term *perspectiva*, which Boethius had chosen as a translation of the Greek *optikē*, was used in the fifteenth century to denote 'the art of representing spatial panoramas or objects graphically on two-dimensional surfaces' [Andersen 2007: xx]. Some four decades after Alberti wrote his works on perspective, Piero della Francesca would use the variant *prospectiva* in his treatise, *De prospectiva pingendi*. Alberti, however, does not use either Latin term, nor either of the corresponding Italian terms (*perspettiva* or *prospettiva*), in *Elements of Painting* and *On Painting*, two texts that are, nonetheless, clearly focused on methods for drawing in perspective.

[3] Alberti's annotated copy of Euclid's *Elements*, a manuscript of Campanus of Novara's translation now housed at St. Mark's library in Venice, is the subject of a recent study [Massalin and Mitrović 2008]; the article's appendix includes transcriptions of all of Alberti's annotations and reproductions of his drawings.

There is also considerable overlap between Alberti's *Elements of Painting* and his own larger treatise *On Painting*, which was also written in two versions, the Italian *Della pittura* and the Latin *De pictura*.[4] The present commentary will therefore explore the relationships between *Elements of Painting*, *On Painting* and Euclid's *Elements*. This will not only lead to a broader understanding of *Elements of Painting* but will also help to clarify a few concepts that are not immediately obvious.

On Painting is widely considered to be one of Alberti's two principal written works, the other being *De re aedificatoria*, one of the most important architectural treatises of the Renaissance and indeed of all time. In *On Painting* Alberti systematically presents, for the first time in history, the geometrical methods for the construction of (linear) perspective, which had been discovered by Filippo Brunelleschi earlier in the century, in the second decade of the Quattrocento.[5] *On Painting* comprises three books. Book I is the most mathematical, offering the definitions of key terms and the principal steps behind the construction of perspective. Book II includes additional construction methods concerning the representation of objects in perspective, as well as some discussion of the virtues of painting. Book III contains more general guidance for the painter and is the least mathematical of the three, though it does include an interesting quote: 'I want the painter, as far as he is able, to be learned in all the liberal arts, but I wish him above all to have a good knowledge of geometry'.[6]

As Alberti writes in his dedication to *On Painting*, Book I, 'which is entirely mathematical, shows how this noble and beautiful art arises from roots within Nature herself'.[7] Alberti starts with the definitions of simple geometrical objects such as point, line, surface, circle, angle, etc., clearly drawing on Euclid for the basic content, but adding much more description pertinent to the viewpoint of a painter. He also makes references, especially in the Latin version, to related facts, topics still in dispute, and stories of the ancients. Thus he exaggerates a bit when he calls Book I 'entirely mathematical,' especially as compared to the first few lists of *Elements of Painting*, as we shall see below.

---

[4] There is some debate as to which version Alberti wrote first, *Della pittura* or *De pictura*, as well as whether Alberti first wrote *Elements of Painting* or *On Painting*. The current prevailing view seems to be that *De pictura* was finished in 1435, *Della pittura* in 1436; see [Greenstein 1992: 235–236, note 1] or [Grafton 2000: 354–355, note 2] for a summary of the scholarly dispute, or for the primary research cf. [Grayson 1998:245–269], [Simonelli 1971] and [Maraschio 1972]. As for the relative chronology of *Elements of Painting* and *On Painting*, it is certain from Alberti's own words in the dedication of *Elementa picturae* that it was written after (at least one of the versions of) *On Painting*. As for *Elementi di pittura*, there is less known about its date relative to *Della pittura* and *De pictura*. Interestingly, the upshot of [Simonelli 1971], though the minority view, puts the chronology as follows: *Elementi di pittura*, *Della pittura*, *De pictura*, and finally *Elementa picturae*.

[5] While earlier attempts at realistic depictions in painting were somewhat successful and included methods now associated with perspective, such as foreshortening and convergence of lines, it is generally accepted that Brunelleschi 'was the first to paint genuine perspective compositions' [Andersen 2007: 11]; and that Alberti was the first to codify the methods of perspective mathematically [Andersen 2007: 17ff]. We refer the interested reader, as well, to the appendix 'On Ancient Roots of Perspective' [Andersen 2007: 723ff], which discusses possible influences of ancient sources such as Euclid's *Optics*.

[6] [Alberti 2004a: 88].

[7] [Alberti 2004a: 35].

In order to describe the construction of perspective, Alberti defines such terms as visual rays (what we now might call lines of sight), outlines (the peripheries or borderlines of objects being viewed), and visual pyramids (composed of the visual rays emanating from the eye and striking the visible surfaces of objects). The real action starts in section 12 (out of 24) of Book I, which ends with the definition of a painting itself:

> Therefore, a painting will be the intersection of a visual pyramid at a given distance, with a fixed centre and certain positions of lights, represented by art with lines and colours on a given surface.[8]

The rest of Book I is devoted to understanding this intersection, with the primary mathematical technique being similar triangles. The upshot is that one learns how to draw a correct perspective of a squared pavement, i.e., a tiled floor, and that one can check the accuracy of the squared pavement with what can be called Alberti's diagonal rule, that a single straight line forms the diagonals of squares that are corner to corner in the pavement (fig. 1).[9]

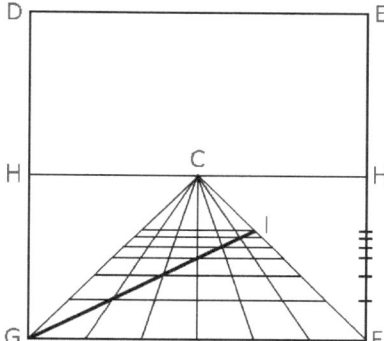

Fig. 1. Alberti's 'diagonal rule' for squared pavements, after [Alberti 2004a: 57, fig. 11]

Moving beyond Book I, we first borrow Martin Kemp's words:

> Book II opens with an extended digression, addressed to 'the young', on the *virtù* of painting, using a cluster of citations from ancient authors to emphasize its 'divine power', moral worth and the high social status of its practitioners.[10]

The bulk of Book II of *On Painting*, however, discusses the roles in painting of three crucial aspects, 'circumscription, composition, and reception of light'; the first of these, and the most important for our purposes, 'is the process of delineating the external outlines on the painting'.[11] Further exploration of circumscription will be more fitting below, as we delve into *Elements of Painting*. For completeness we note that Book III of *On Painting*, as with the beginning of Book II, does not help inform our study of Alberti's *Elements of*

---

[8] [Alberti 2004a: 48].
[9] [Alberti 2004a: 58].
[10] [Alberti 2004a: 14].
[11] [Alberti 2004a: 65].

Painting, which is much shorter and focuses only on essential drawing skills necessary for the painter to execute a perspective.

Proceeding to a discussion of the *Elements of Painting*, then, brings us first back to Book I of *On Painting* and to Euclid's *Elements*. As in Book I of *On Painting*, Alberti begins *Elements of Painting* by working from first principles; the difference is that in the latter he is much less wordy, so that his writing resembles Euclid's to a much greater extent. For example, Alberti's definition of **point** in A.1 ('The point is said to be that which cannot be divided at all into any parts') is essentially the same as the very first definition of *Elements* ('A **point** is that which has no part' [Euclid 1956: vol. 1, 153 (I, Def. 1)]). He does not just slavishly copy Euclid's definitions, however. His next definition, of **line** in A.2 ('A line is said to be almost <like>[12] a point stretched out in length. Thus the longitude <length> of a line can be divided, <but> not its latitude <width>'), combines and reworks the next two definitions of Euclid's:

2. A **line** is breadthless length.

3. The extremities of a line are points.[13]

In a similar way, Alberti's definition of **surface** in A.3 is a combination of Euclid's I, Def. 5 and 6 (as we shall see below, Alberti saves I, Def. 4 for his D.5). Alberti ends List A with the definition of **body**, which corresponds to Euclid's definition of the synonymous term 'solid' in a much later book of *Elements* (XI, Def. 1).[14]

List B starts with Alberti noting that the contents of List A 'were stated by the ancients'.[15] Similarly we learn, at the beginning of List C, that List B's 'considerations are ours'. What is interesting is that the terms of List B are the basically the same terms from List A, presented in reverse order and reformulated so as to pertain more to how a painter sees. The (re)definition of **body** in B.1 is the only one that can be seen as similar to Euclid, namely XI, Def. 2. The rest really are 'ours' (i.e., Alberti's). The idea behind B.2 and B.3 is that a **surface** is the visible part of a body, and that a surface is bounded by an **edge**, a term which has taken the place of 'line' from List A (and from Euclid). Alberti also introduces, in B.3 and B.4, a term borrowed from Latin, namely **_discrimen_**, the boundary

---

[12] Integrations to the text for the sake of clarity are added angle brackets < >; integrations included in the 1973 Grayson edition are added in { } brackets.

[13] [Euclid 1956: vol. 1, 153 (I, Defs. 2 and 3)]. Fibonacci (Leonardo of Pisa) more faithfully combines Euclid's I, Def. 2 and 3 into one when he offers 'Line is length lacking width, with points as its ends' in *De practica geometrie* and [Hughes 2008: 5]. This is pertinent since Alberti was well aware of *De practica geometrie*; see our commentary to Problem 12 of *Ludi matematici*, pp. 101–106 in this present volume.

[14] [Euclid 1956: vol. 3, 260]; as Heath notes on p. 262, 'body' was the term used by Aristotle, so perhaps this was Alberti's reason for choosing it despite the fact that 'solid' was used by Euclid.

[15] Our interpretation of Alberti's words differs from that of Golsenne et al. [Alberti 2004b] in this case. We translate Alberti's Italian *Queste dissero li antique* (literally, 'These said the ancients') as 'These [statements above] were stated by the ancients'. The French translators render it *Les Anciens ont donné les définitions suivantes* ('The Ancients gave the following definitions') [Alberti 2004b: 280], which does not make sense based on the content. Moreover, from later lists it is clear that Alberti often starts a new list with comments on the preceding one. Also note the different labeling of lists in [Alberti 2004b]; what we call List A through List D they group into one, so our E corresponds to their B, our F to their C, etc. Our grouping matches that chosen by Grayson.

between two surfaces. To give Alberti's example in B.4 some context, think of someone's head with the hair parted in the middle. The part itself is what Alberti is giving as his example of a *discrimen*. The issue here is that a painter must repeatedly choose where one area of colour will end and the next will begin, and the border between these two is what Alberti is calling *discrimen*. (The fact of the matter is that Alberti does not use *discrimen* from List C onwards, so it is curious that he takes the time to define the term with a fair amount of detail and then avoids using it.) The final item of the second list, B.5, states that a **point** occurs at the intersection of edges.

Albert is still not done with these terms, which he revisits yet again in List C, back in the original order of List A. He states that the third list is for 'our painters', and we soon realize this means not only what a painter sees but also what a painter actually draws on the canvas. This is in keeping with the viewpoint that Alberti takes in his one-page note *De punctis et lineis apud pictores* (*On the Points and Lines Used by Painters*), which contrasts the theoretical constructs dictated by mathematics with the real-world constraints of the painter.[16] In C.1 a **point** is now redefined as the smallest of small marks. Here Alberti is straying far from the rigor of Euclid, into the realm of self-contradiction, since no mark is so small that 'nothing can be smaller'. Rather than accusing Alberti of being non-rigorous, however, we should note that what he is apparently trying to do is bring the abstract notions of Euclid into the more concrete realm of painting, which is fitting for his intended readership. In the same vein, a **line** in C.2 is a 'very fine inscription'. The Italian version also includes **edge** in C.3, repeating the concept of the outline of a surface. Finally, in C.4 Alberti introduces **area** as the space on the canvas corresponding to the surface viewed. It is important to realize that when Alberti refers to drawing an area later in *Elements of Painting*, he really means drawing the outline or edge of an area, which corresponds to circumscription, one of Alberti's three crucial aspects of painting we discussed above. The kind of outline Alberti deals with in *Elements of Painting* is what mathematicians now call a 'simple closed curve' occurring in a two dimensional plane, where the term 'curve' in this context allows for straight line segments as well as curves, 'closed' means the whole border is encircled, and 'simple' means that it does not cross itself – objects such as circles, polygons (regular or irregular), semicircles, etc.

Alberti starts the fourth list of *Elements of Painting* with two non-Euclidean terms. To understand these terms it will help to go back to our discussion of *On Painting*, where we had left off in Book II under the topic of circumscription. Alberti describes an aid he developed in order to draw outlines realistically, the use of a 'veil', a translucent piece of cloth having sewn into it thicker threads in the form of a grid, which corresponds to an analogous grid on the painting panel (fig. 2).[17]

---

[16] See [Alberti 1890: 66]. Mancini attributes this one-page note to Alberti, stating that if follows a manuscript of *De pictura* (*Ex codicibus Taurinensi 1184 in bibliotheca regii Studii, f° 60, et Lucensi 1448 in bibliotheca Regia, f° 52. — Nota haec invenitur post libros* De pictura).
[17] [Alberti 2004a: 67].

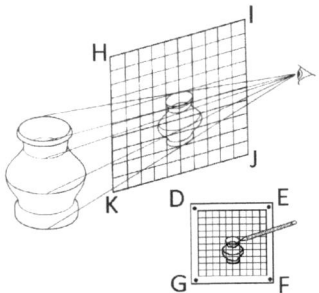

Fig. 2. After [Alberti 2004a: 66, fig. 12]

An object seen through the veil, inherently in perspective, can easily be drawn on the painting panel by noting where its outline lies in relation to the grid on the veil. Two sections later, Alberti instructs that one can draw a circle in perspective by first drawing a true circle on a grid, drawing the grid in perspective, and then transferring the segments of the circle in each square of the grid onto the corresponding squares of the grid in perspective; more on this specific construction below (fig. 3).

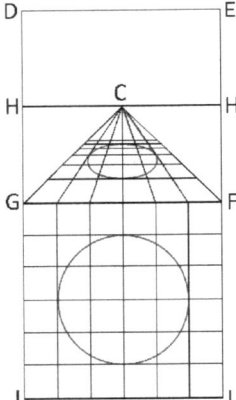

Fig. 3. After [Alberti 2004a: 70, fig. 14]

Note how we use the terms 'true circle' and 'circle in perspective' in the discussion above. In *Elements of Painting* Alberti apparently felt the need to define different adjectives to distinguish between objects seen straight on and objects seen at a tilt, and hence the definitions for **concentric** in D.1 and **comminuted** in D.2, respectively. One can relate Alberti's use of 'concentric' to the standard modern usage (having a common centre, having a common axis) by realizing that, in order to view the true shape of an area, one's eye must be centred on the area, i.e., the eye must be positioned on the axis passing through the centre of the area and perpendicular to it. (This is easiest to visualize with an area that has an obvious centre, such as a circle, square, regular polygon, etc.) As for the second term, 'comminuted' is roughly synonymous with the modern term 'foreshortened', though it is quite common just to use the phrase 'in perspective' (despite the fact that some objects drawn in perspective, namely those objects that are parallel to the picture plane, can be true in shape).

Alberti goes back to Euclid for the next definition, that of **proportional** in D.3. This term is defined twice in the *Elements*, in V, Def. 6 and VII, Def. 20, with the latter being the closer match.[18] While Alberti's definitions of proportional and concentric may appear redundant in a first reading, note that concentric, as with comminuted, refers more to the nature of an area depending on the viewpoint of the painter, while proportional refers more to the relationships between areas. In fact, it is just as possible for two comminuted areas to be proportional as it is for two concentric areas.

Alberti strays from Euclid once again with what is undoubtedly the least obvious definition in *Elements of Painting*, **commensurate point**, defined in D.4 as follows:

> A commensurate point will be that which will have coequal distances from the other points, whether they are of concentric or comminuted [areas].

The Latin version has a different formulation:

> A commensurate point in a picture will be one with distances in a certain relationship to other points in the picture.[19]

Surely Alberti could have written a Latin version closer to the Italian version of this definition, if he had intended to do so; thus he apparently felt a need to reformulate the definition of commensurate point. Perhaps he found fault with his original definition, which he then sought to correct in the Latin. An analysis of the Italian definition makes this appear likely.

First note that in our translation of Alberti's Italian *averà sue coequali distanze*, we do not choose 'will have equal distances' (nor simply 'will be equidistant') but rather 'will have coequal distances'.[20] Admittedly, what exactly Alberti means by 'coequal' is not clear, but we still maintain that 'equal' is not the logical choice, unless the translator assumes either a serious manuscript transcription error (which can never be ruled out), or the possibility that *coequali* and *equali* were the same either to Alberti or to a later transcriber, or some sort of error by Alberti himself. Moreover, if Alberti himself had meant the strict notion of 'equal' rather than the presumably more flexible 'coequal', then he could not have meant 'the other points' strictly: if one is strict both with 'equidistant' and with 'the other points' (meaning <u>all</u> the other points), then there is only one area that satisfies the definition: a circle.[21] Indeed, the centre of a circle would be a commensurate

---

[18] [Euclid 1956: vol. 2, 114 and 278]. The latter also avoids potential confusion, since the former depends on V Def. 5, which is notorious for having confounded readers for centuries, despite being mathematically correct (and quite an advanced treatment of a surprisingly difficult topic); see Heath's extended note to V Def. 5 [Euclid 1956: vol. 2, 120–129].

[19] Cf. Alberti's Latin version: *Punctum commensuratum in pictura erit cum a caeteris picturae punctis certa intervalli ratione distabit* (trans. Stephen R. Wassell).

[20] This is the opposite choice to the French translators: *sera placé à égale distance* ('will be placed at equal distance') [Alberti 2004b: 283].

[21] Recall that by 'area' we really mean the area's edge, as Alberti himself does.

point in the strictest sense, since it is equidistant from all the points of the circle.[22] Moreover, no other area besides a circle can be said to have a commensurate point, in the strictest sense, because no point could have the same distance from all the points of the area. Obviously it would make no sense to assume that Alberti meant his definition to pertain only to circles. Thus, since either *coequali* or 'the other points' (or both) had to be loosely interpreted, this might have given Alberti the impetus to change the definition in the Latin version.

In the Latin version we still have to face the fact that what is meant by 'other points' is vague; in fact, Alberti does not even maintain the clarification that the 'other points' come from either a concentric area or a comminuted area (not that this is much of a clarification). And while the issue of the interpretation of *coequali* becomes moot in the Latin version, it seems that Alberti injects at least as much inherent looseness of interpretation into the Latin.

Because of this we end up having to try to glean information from the ways in which the term 'commensurate point' is used in *Elements of Painting*, as well as how it might relate to *On Painting*. The issue is that the use of 'commensurate point' in *Elements of Painting* is as vague as the definition itself; the reader is instructed to draw a commensurate point inside an area, or outside an area, relative to both concentric and comminuted areas, and there is never any specific instruction as to how a commensurate point is to be determined. Perhaps the construction described above from *On Painting*, of a circle in perspective (i.e., a comminuted circle), is instructive. When discussing how he transfers the true circle on the grid to the comminuted circle in the painting, Alberti states:

> But as it would be an immense labour to cut the whole circle at many places with an almost infinite number of small parallels until the outline of the circle were continuously marked with a numerous succession of points, when I have noted eight or some other suitable number of intersections, I use my judgement to set down the circumference of the circle in the painting in accordance with these indications.[23]

---

[22] A tempting but fruitless avenue of logic is based on the fact that it is not just the centre of a circle that can be a commensurate point with this interpretation; so, too, can any point on the axis passing through the centre of the circle and perpendicular to the plane containing the circle. And this leads us back to the discussion of concentric versus comminuted, for as noted above, an eye positioned on this axis will lead to a view of the circle as concentric. By allowing flexibility with *coequali* or 'the other points' (or both), one could make such a statement for squares, regular polygons, and the like. Thus, in this interpretation, the commensurate points of an area are precisely those which lead to a concentric view of the area. Locating such an eye position could be useful for visualizing areas that need to be painted, but as it turns out, this is not what Alberti seems to have in mind with the exercises that involve commensurate points. Indeed, the problem with this interpretation is that Alberti's exercises include instructions for drawing a commensurate point outside of an area, so interpreting a commensurate point outside of an area as an eye position is not altogether compelling. On a related note, we have considered how 'commensurate point' might relate to what Andersen calls 'an *Alberti point* – the characteristic point of an Alberti construction' (defined in the caption to Figure II.18 [Andersen 2007: 41]), but this also seems fruitless.

[23] [Alberti 2004a: 71].

Is it possible that the eight points described here could be called eight commensurate points? Could a commensurate point simply be a construction point in a perspective drawing used to find the position of a line, curve, edge, etc.? Should we read any more constraints into the definitions of commensurate point in the Italian and Latin versions of *Elements of Painting*?

One way to investigate these questions is to study the Quattrocento treatises that were influenced by Alberti's works on perspective. The most important such treatise is Piero della Francesca's *On the perspective of painting* (*De prospectiva pingendi*, or *De prospectiva* for short), since it advanced the emerging discipline of perspective considerably, providing descriptions of four different perspective techniques, the first of which is equivalent to the method that Alberti gives in *On Painting*.[24] Because it is so advanced relative to Alberti's text, however, Piero's is not as helpful with regard to our present queries. A more helpful text in this regard is Filarete's *Treatise on architecture* (*Trattato di architettura*), which was closer to Alberti both chronologically and in content.[25] Because of this, Alessandro Gambuti analyzes how Filarete's tract relates to Alberti's *On Painting* and *Elements of Painting*, and he starts with two observations: that Alberti's two treatises are, for Filarete, the only and the authentic foundation for rendering reality in painting, and that Filarete himself comments on how Alberti's two treatises are interrelated.

Taking Gambuti's lead to analyze Filarete's tract, in particular Books 22 through 24, we see how the latter draws heavily from Alberti for the wording of such concepts as point, line, surface, area and edge. We see that Gambuti and Filarete are on the same page as we are with regard to the meaning of concentric, comminuted, and proportional areas. And then we arrive at commensurate point, bringing us back to the story at hand. Gambuti suggests that Filarete's 'given point' (*punto dato*) may correspond to Alberti's commensurate point (*punto commensurato*), but he laments that Filarete gives 'no adequate explanation such as might be expected' of the 'given point'.[26] Indeed, Filarete

---

[24] [Andersen 2007: 17 and 34ff].

[25] [Andersen 2007: 43]; Filarete is the pseudonym by which Antonio di Pietro Averlino (ca. 1400–1469) is known.

[26] Cf. [Gambuti 1972: 168]: *Ma del* punto dato *(che corrisponde verosimilmente al* punto commensurato *degli* Elementi, *definito come* quello qual da li altri puncti o vi da la concentrica o vòi da la comminuta haverà sue coequali distantie*) non viene data un'adeguata spiegazione, come ci si aspetterebbe* (But of the *punto dato* [which probably corresponds to the *punto commensurato* of the *Elements*, defined as *that which will have coequal distances from the other points, whether they are of concentric or comminuted* <areas>] no adequate explanation such as might be expected is given). As Gambuti points out, Filarete refers the reader to Alberti for more on the theory, since Filarete is more interested in putting theory into practice; cf. [Filarete 1972: 642]: *E tutto ciò che 'l disegno contiene se diriva da questo punto dato, e dalla linea, e dalli angoli. Ora è mestiero a 'ntendere e mettergli in uso. E in pratica molte sottilità ci si può dire in queste linee, e punti, e superfice, e angoli, e corpo, e aere, e comminute e proporzionali e centriche, come che per lo 'detto' Battista è scritto, e non così è trattato, ma perché più facile ti sia a dovere intendere e non così tante ne diremo* (And all that the drawing contains derives from this given point and from the line, and from the angles. Now it is our task to understand and put them into use. And in practice many subtleties can be said about these lines, points, surfaces, angles, bod[ies], areas, [whether the areas be] comminuted, proportional, or [con]centric, as has been written by the 'so-called' Battista, and has not

does not elaborate on the term *punto dato*, and so it is impossible to judge whether Gambuti's association of it with Alberti's commensurate point is correct. We shall consider Filarete's illustrations below when we analyze the exercises that involve commensurate points.

After the saga concerning Alberti's commensurate point, the remaining definitions of *Elements of Painting* are anticlimactic. They are readily understood, and several relate to Euclid. The definition of **straight line** in D.5 ('A straight line is that stroke drawn from one point to another by the shortest way, and which does not curve in any of the spaces') differs completely from *Elements* I, Def. 5 ('A straight line is a line which lies evenly with the points on itself' [Euclid 1956: vol. 1, 153]); Heath calls Euclid's formulation 'hopelessly obscure' and lists several other ancient Greek sources, some with variations on the 'shortest distance' theme evident in D.5.[27] Alberti offers two definitions in D.6, **rectilinear** and **curvilinear**, the definition of rectilinear being very close to part of Euclid I, Def. 19. The definition of curvilinear depends on Alberti's definition of **curved line** in D.7, which itself is problematic since it implicitly assumes that the only source for curves can be pieces of circles.[28] In defining a **circle** in D.8, Alberti complicates matters at first but then ends up with language analogous to Euclid I, Def. 15 for circle and I, Def. 16 for centre. The definitions of *Elements of Painting* end with D.9, which includes definitions for **angle** (corresponding to Euclid's 'plane angle' of I, Def. 8), **right angle** (corresponding to I, Def. 10), **non-right angle**, **rectangle** (corresponding to Euclid's 'oblong' of I, Def. 22), **non-rectangle**, **triangle** and **quadrangle** (the last two corresponding to Euclid's 'trilateral' and 'quadrilateral' of I, Def. 19).

Alberti starts List E by stating that '[t]he definitions given up to this point are sufficient' and that the remainder of the treatise will set out in order (from the basic to the complex) the skills necessary for the practitioner to understand how to draw (meaning, implicitly, how to draw in perspective). Indeed, the rest of the treatise comprises lists of drawing exercises, and List E is basic enough that several items relate directly to Euclid's *Elements*. Not surprisingly, Alberti now draws from Euclid's postulates and propositions rather than definitions. For example, E.1 corresponds to the first of Euclid's postulates (I, Post. 1). Neither E.2 nor E.3 has a direct correspondence to Euclid, but E.3 is certainly related to I, Post. 2, while E.2 relates to the many instances in the *Elements* where line segments are divided at certain points. Next we find relationships to later postulates: E.4 (where 'equidistant from' means 'parallel to') is very similar to I, Prop. 31, and E.5 is very similar to I, Prop. 11. The next two, E.6 and E.7, together imply that the painter should be able to reproduce any triangle, right or non-right, and this is fitting given the importance of congruence of triangles in Euclid's *Elements*. Exercise E.8, the last before Alberti involves a commensurate point, requires that the practitioner know how to draw any concentric area, and the obvious underlying issue is that comminuted areas are more easily constructed if first drawn as concentric.

---

been dealt with this way, so that it be easier for you to understand[,] we will not say so much about them). Trans. Stephen R. Wassell.

[27] See [Euclid 1956: vol. 1, 165–167].

[28] It could be argued that this is true in some sort of infinitesimal way, but we have no reason to believe that Alberti was thinking as such.

The flavour of the exercises is about to change at this point, but let us first consider that Gambuti provides illustrations adapted from the exercises, with the following stated intent:

> In reconsidering the Albertian text, the attempt was made to give it a graphic interpretation for the first time, in accordance with the Latin formulation of the problems. In general we tried to adapt ourselves to Renaissance perspectives, whose example permitted us – we hope – to conclude our experiment with an outcome that is not altogether imperfect.[29]

Gambuti succeeds in not being too imperfect, and his figures are helpful and thought-provoking in many ways. Given the very nature of these interpretative illustrations, however, one must view them with a questioning eye. Those corresponding to E1–E8 are quite straightforward, given the content of the exercises. In some cases one could even argue that a given illustration, while being only an example of its exercise, in some way captures the essence of the exercise in general. No claims of this are made, however, and so it is best for all interested parties to assume that each illustration is meant to give just one example of its exercise.

Finally, in E.9 and E.10, we arrive at the first mention of commensurate point in the exercises and at the moment in *Elements of Painting* where we start to see the true flavour of the majority of the exercises:

> E.9. Inscribing a commensurate point in a rectilinear and concentric area.
>
> E.10. Placing in one drawn rectilinear and concentric area another smaller concentric and rectilinear [area].

We group these two not only because they seem to be two steps towards a single goal but also because Gambuti groups these together for his Figure 12, which he states was inspired by an illustration in Filarete's tract on architecture (fig. 4).

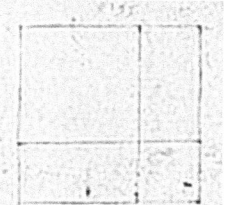 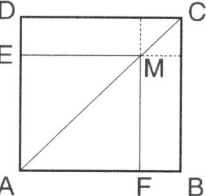

Fig. 4. a, left) Illustration from Filarete's *Trattato di architettura* (ca. 1465), Bk. 22, fol. 58v from Ms. Magliabechiano II.I.140, fol. 174, Biblioteca Nazionale Centrale di Firenze; b, right) after [Gambuti 1972: 144, fig. 12]

---

[29] Cf. [Gambuti 1972: 137]: *Nel ristudiare il testo alberitano, si è tentato per la prima volta di darne un'interpretazione grafica, seguendo la formulazione latina dei problem. In generale, abbiamo cercato di adeguarci ai prospettivi rinascimentali, il cui esempio ci ha consentito — speriamo — di condurre l'esperimento ad un esito non del tutto imperfecto* (trans. Kim Williams). Note that Gambuti's illustrations are based on the exercises from the Latin version, which do differ somewhat from the Italian version. Golsenne et al. [Alberti 2004b] present their own versions of many of these illustrations; while this is helpful in making Gambuti's interpretations available to a wider readership, one wonders why there is no discussion offered, except for one sentence giving credit to Gambuti for aid.

Points labelled *M* by Gambuti are commensurate points.[30] It is reasonable to accept that *M* satisfies the definition of commensurate point in Alberti's Italian version with suitably non-strict interpretations of either 'coequal' or 'the other points'; similarly, since Alberti's Latin definition seems to be even looser than his Italian, Gambuti's point *M* may very well be a reasonable choice for a commensurate point. If we do accept this, then it stands to reason that there are an infinite number of other possible choices for *M* on the diagonals alone, not to mention other possible placements for commensurate points. Already we see that Gambuti seems to view the concept of commensurate point as a fairly flexible one.[31] This would agree with our analysis of Alberti's wording.

It is instructive to realize in advance that the last two exercises in *Elements of Painting* are:

L.13. Drawing any area in any other area.

L.14. In proximity to any area, drawing another area similar to whatever kind of surface viewed.

Apart from the exercises that relate to Euclid, most of Alberti's exercises are geared towards examples of L.13, with various kinds of areas (rectilinear, curvilinear, and circular) and viewpoints (concentric and comminuted). Interestingly, there are not many examples of L.14 in the treatise (most areas are one within another rather than side by side), and perhaps this last exercise was added by Alberti as a sort of catchall, to show that he realized painting requires much flexibility.

In any case, the exercises geared towards examples of L.13 often come in pairs, similar to E.9 and E.10 above, so that the first involves placing a commensurate point inside (or outside) a given area, and the second involves drawing a second area inside (or outside) the given area. In fact E.11 and E.12 are just like E.9 and E.10, except that the new rectilinear area is now outside the given one. Exercises E.13 and E.14 can also be grouped, but in a different way. There is no commensurate point mentioned, nor a specification that one area is inside another. The next pair, E.15 and E.16, is back to 'usual' form like E.9 and E.10, while E.17 is the second of a usual pair with the first missing. And finishing List E, E.18 can be grouped with E.15 and E.16, while E.19 can be grouped with E.17. The upshot of List E is that the practitioner should now be able to draw concentric rectilinear areas (i.e., polygons true to shape), proportional to one another, inside or outside one another, side by side – essentially, in any configuration.

A quick observation of terminology is that Alberti has introduced the term 'angular' in E.16 through E.19, presumably as a synonym for rectilinear in these exercises. Later, at the start of List I, he refers to 'angular rectilinear' and 'angular curvilinear' areas, in which it seems that 'angular' is a bit of a throw-away word, though perhaps stressing that he is referring to general areas and not, for example, ones with right angles only. Then in Lists I

---

[30] This is made explicit in the captions to Figures 26 and 27 [Gambuti 1972: 150], and it is clearly intended throughout.

[31] The other possibility is that Gambuti viewed his Figure 12 as somehow the only interpretation of Alberti's E9 and E10, but this does not seem to be the case. If it were, we would have to disagree, since Alberti's wording is quite broad.

and L Alberti uses 'angular' in its original way, presumably as a synonym for rectilinear again.

At the beginning of List F, Alberti states that List E suffices for rectilinear areas and that List F will concern curvilinear ones, which for Alberti means areas enclosed by pieces of circles (and presumably straight line segments as also allowed by D.6). While F.1 through F.8 do not mention commensurate points, the remaining exercises F.9 through F.17 are then heavily focused on commensurate points. Together, F.1 through F.8 require the ability to draw any curvilinear area true to shape, and to reproduce such at the same scale or proportionally smaller or larger. Gambuti offers an interesting illustrative choice for F.5, namely a rendition of the figure included in Mancini's 1890 edition of *De lunularum quadratura* (fig. 5).[32]

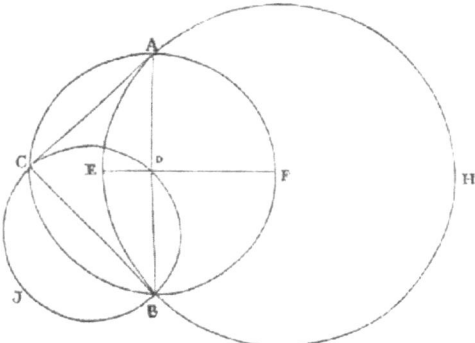

Fig. 5. The figure attributed to mathematician Francesco Siacci that appeared in Mancini's 1890 edition of *De lunularum quadratura* [Alberti 1890: 306]

While the words of F.5 could be construed to include the squaring of the lune construction, the question arises as to how often lunes appear in perspective drawings. We view F.5 and F.6 as dealing with the ability to draw areas in painting corresponding to objects which have other objects placed in front of them. In any case, the rest of the exercises deal with inside/outside relationships, with F.9 and F.10 being one usual pairing (place commensurate point, draw area), F.11 and F12 another. Four of the last five exercises deal only with marking/placing commensurate points, with F.14 being the exception. One wonders if Alberti really meant F.14 to open the door to general concentric areas, or if a word error was made at some stage.

List G starts by saying goodbye to concentric curvilinear areas, and at the beginning of List I (there is no List H, J or K), Alberti tells us that List G is for concentric circles. First there are some basic exercises, with G.2 essentially the same as Euclid III, Prop. 1, and G.3 equivalent to I, Post. 3.[33] Gambuti's illustration of G.2 is one of the most successful in his publication from the point of view of demonstrating a general principal (fig. 6).[34]

---

[32] See our translation and commentary of *On Squaring the Lune*, p. 203ff in this present volume.

[33] [Euclid 1956: vol. 2, 7; vol. 1, 154]; Book III, Prop. 1 does not include the diameter in the statement of the proposition, but the diameter is found in the proof along with the centre.

[34] [Gambuti 1972: 151].

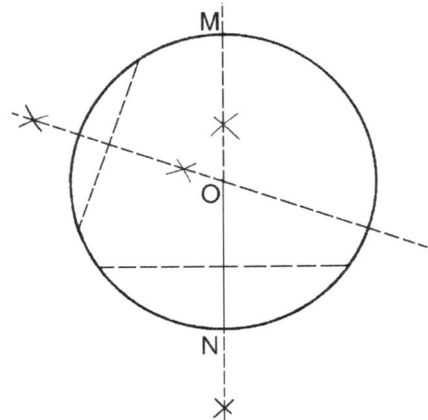

Fig. 6. After [Gambuti 1972: 151, fig. 29]

The rest of the exercises from G.4 to G.14 are dominated by placing commensurate points and drawing concentric circles proportional to other circles, usually either inside or outside. (Of course, every circle is proportional to any other, so it is a bit redundant to specify this.) One naturally wonders whether or not Alberti meant concentric in the modern sense at all, at least in List G, but he never specifies that the circles should share the same centre. All of Gambuti's drawings for G.4 onwards involve concentric (in the modern sense) circles, and his illustration covering exercises G.6 through G.9 deserves comment (fig. 7); Gambuti has chosen the commensurate points M and N based on a particular method, as he often does, despite the fact that no guidance is given by Alberti.[35]

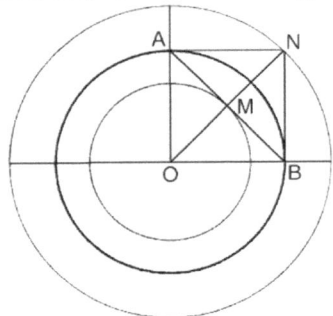

Fig. 7. After [Gambuti 1972: 152, fig. 32]

At the beginning of List I, we learn that the rest of *Elements of Painting* will be about comminuted areas, starting with rectilinear in List I. It is interesting that Alberti would start right away, in I.1 through I.3, by mixing comminuted areas with concentric, one inside the other. Gambuti's illustration of I.2 is noteworthy in that the same rectilinear shape (an irregular pentagon) is used for both the concentric and comminuted areas, and

---

[35] [Gambuti 1972: 152]; to be precise, Gambuti bases his illustrations on the Latin version, so for him this illustration corresponds to exercises 6 through 8 of this list.

incidentally it shows a few construction points used to draw the comminuted irregular pentagon (fig. 8).

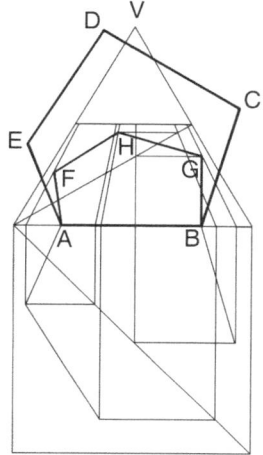

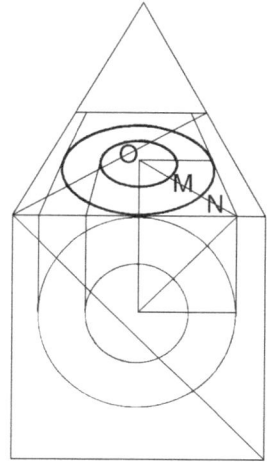

Fig. 8. After [Gambuti 1972: 153, fig. 35]   Fig. 9. After [Gambuti 1972: 159, fig. 47]

Could such points be what Alberti meant by commensurate points? In I.3, we do not know exactly what Alberti means by circumscribing, for that matter. Recall that 'circumscription' is one of the three crucial aspects of painting in *On Painting*, so perhaps 'to circumscribe' for Alberti is simply to delineate the outline. If this is true, then we should not restrict circumscribing to the modern sense, where the outer figure touches the inner figure in a minimal way at some number of points. Exercises I.4 through I.12 deal with comminuted areas only, and they can be grouped into usual pairs or triplets

At the beginning of List L, we see that Alberti himself may have been aware of the repetitive nature of the last few lists of exercises. Having just finished the rectilinear comminuted areas and faced with both curvilinear comminuted and circular comminuted, Alberti skips over the former by claiming that curvilinear comminuted areas are covered in the same basic way as the rectilinear comminuted areas. Fair enough: this makes List L on circular comminuted areas the last. We have already discussed above the catchall exercises that Alberti ends with, L.13 and L.14. Leading up to these, as one might expect, exercises L.1 through L.12 involve a potpourri of areas. Of these twelve, each exercise except for L.4 involves at least one circular area, each except for L.3 involves only comminuted areas, and four exercises involve commensurate points. Gambuti's final illustration, corresponding fairly closely to our L.10 through L.12, indeed seems like a good example of the kinds of skills that Alberti would find important for drawing in perspective (fig. 9).

Finally, Alberti ends with section M, which is not a list but simply a concluding set of statements that requires no commentary.

It would be false to conclude that we understand exactly what Alberti meant by a commensurate point. What seems to be suggested by our analyses, however, is that a commensurate point is some sort of construction point used in the process of drawing in perspective. It may be a point of symmetry of an area, or it may be a point of intersection occurring when drawing a grid, whether the grid is viewed straight on (concentric) or foreshortened (comminuted).[36] In any case, it is quite possible that Alberti was ahead of his time. Kirsti Andersen writes:

> Mathematicians in later times, starting with Simon Stevin, realized that the cardinal problem in perspective is to construct the image of an arbitrary point – the image of a polygon being determined by the images of its vertices.[37]

Perhaps Alberti's focus on commensurate points shows that he was inherently aware of this, at least subconsciously. We may never know exactly what Alberti meant with his definition of commensurate point, Italian or Latin, but it would not be surprising in the least that Alberti possessed some keen insight and foresight in this case.

---

[36] It is instructive, in this regard, to quote what Richard Schofield produced when we asked him for an independent translation of commensurate point from the Latin: 'A commensurate point in a picture will be one with distances from other points in the picture that can be precisely calculated'.

[37] [Andersen 2007: 44]; Stevin (1548–1620) lived over a century later than Alberti.

On Writing in Ciphers

K. Williams et al. (eds.), *The Mathematical Works of Leon Battista Alberti*,
DOI 10.1007/978-3-0346-0474-1_4, © Springer Basel AG 2010

I

Those who are charged with the highest affairs know by experience how important it is to have a very trustworthy person with whom to reveal projects and decisions of the most secret nature, without ever having reasons to regret it.[1] This is rare, because men so often tend to perfidy, and so secure systems called *ciphers* have been devised, which would be useful, except for those who manage, through art and ingenuity, to read and interpret them. I admit that those in power may be served by those who are expert in such operations, if by their means the schemes and machinations of enemies can be revealed, but in my opinion, it is even more advantageous to be able to communicate our intentions to another, no matter how far away, in a way that no other mortal except the intended recipient of the missive is able to read them.

This present booklet of mine is a thorough treatment of both aspects of the question, paving the way for and directing the investigation into other people's secrets, as well as how, as we shall see, to protect secrets of your own.

The present circumstances persuaded me to send these notes of mine to you. Many friends who are devoted to you exhorted me to do so. I will be delighted if this work pleases you.[i]

II

While I was with Dati in the papal gardens in the Vatican, discussing, as usual, literary matters, we happened to praise in the most enthusiastic terms the German inventor who has recently made it possible, by means of a system of moveable type, to reproduce from a single exemplar more than two hundred volumes in one hundred days with the help of no more than three men. This makes it possible to obtain an entire, large format page from a single impression. That having brought us to similar appreciations of the brilliant discoveries in the most diverse areas, Dati fervidly admired those who, faced with strange characters with unusual meanings known only to the writers and receivers, called ciphers, have the skill to read and interpret them.[ii]

Said Dati, looking at me, 'You, who have always been involved in investigating occult arts and investigating the mysteries of nature, what do you think about these interpreters of ciphers and revealers of secrets, if we can call them that? Have you any knowledge of this subject?'

I said, smiling, 'It is probably you, chief secretary to the Pope, who has sometimes had to make use of coded writing in dealing with those political affairs that must be kept top secret'.

'That's so', answered Dati, 'and in my position I wouldn't mind being able to deal with them myself, without having to turn to an outside interpreter. Such messages are sometimes intercepted by spies and there arrive coded messages that should not be overlooked. If then you have discovered something in this regard, I pray you to let me in on it'.[iii]

---

[1] Two kinds of notes appear in this text. Translator's notes are numbered in Arabic numerals and appear as footnotes; commentary notes are numbered in Roman numerals and appear in the commentary which follows the translation.

I thus promised to put my mind to the matter, as far as I was able, so that this request of his would not go unanswered. Thus I did, and when it seemed to me that, by means of investigation and reflection, I had mostly accomplished the task I had taken on, I set down in these notes the results of my research, both to comply with Dati's wishes and to present to my fond admirers, as I do from time to time, a new work that can be added to the number of my works. I will tell how I proceeded, as it will be helpful in what follows.

III

In order to investigate more thoroughly, I began by asking myself precisely in what ciphers consist.

It thus came to my mind to me to define cipher as writing in which signs arbitrarily assume a significance that is decided on by the writers and which is unintelligible to anybody else. This said, it was necessary to establish two criteria: one is that each of the correspondents had to follow a consistent and precise rule, which would make it possible to understand reciprocally from the written text what advice, question, information, and so forth was intended

Second, the method of writing adopted had not only to be uncommon and never before seen, but also impossible to decipher by the most skilful investigators and the most acute interpreters. Without a doubt, if my investigations have been sufficiently deep, both of these aspects depend on the way in which we habitually use characters in writing. Having ascertained this, it seemed opportune to me to consider first of all, how the letters function in writing, and then on which principles writing is based on and executed.

I went about this with not indifferent industry and care, reflecting again and again on the elements of writing, and investigating intensely until I had clear in mind some fundamental concepts, which now the most brilliant minds will acknowledge as having contributed to understanding the whole question of ciphers.

First I will set forth the results of these reflections of mine. Although they appear to regard the techniques of deciphering the codes of others, in reality they furnish many suggestions that can be followed in order to devise a coded writing that is unintelligible to the experts. After that I will deal with the diverse ways, both common and obscure, of creating codes that are functional and suitable for use, which will be enthusiastically appreciated by those who are knowledgeable in the field. Finally, I will present a coded writing of my own invention: I am sure that, once you have understood it, you will appreciate it and congratulate me. So now let us begin.[iv]

IV

As I reflected what became exceedingly clear to me was the concept that all of our verbal and written expressions are constituted of words, and that the words themselves can be broken down into syllables that are in their turn formed of letters; it thus seemed to me important to reveal their intrinsic functioning with respect to writing as well as the differences that exist between the letters. I could go on at great length about this, but as briefly as possible, just let me say that the arrangement and the number of letters constitute various syllables which, when combined, form finished words, which differ in sound and meaning. First I shall speak of number and how this is connected to numeric ratios.[2]

This phenomenon above all involves the vowels. We will begin with them. A syllable is constituted of either a single vowel, a consonant associated to a vowel, or several consonants united to a vowel. Without vowels there are no syllables. Thus, if we take, for example, one or two pages of poetry or prose and extract the vowels and consonants, listing them in separate series, vowels on one side and consonants on the other, you will no doubt find that there are numerous vowels.

From my calculations, it turns out that in the case of poetry, the number of consonants exceeds the number of vowels by no more than an octave,[3] while in the case of prose the consonants do not usually exceed the vowels by a ratio greater than a sesquialtera.[4] If in fact we add up all the vowels on a page, let's say there are three hundred, the overall sum of the consonants will be four hundred. I further noticed that, among the vowels, the letter O, not only occurs fewer times than the consonants, but also than the other vowels.

Almost as rare is the vowel A. Rare also is the letter U used as a vowel, while more frequent is the letter V used as a consonant.[5] Elsewhere, where I dealt with questions of a linguistic or grammatical nature, I had counselled writing it with a curved bar almost like a ꝟ, its sound being somewhat intermediate between B and V, and among the ancient writers were those who thought to write it like an upside down F, thus: ꟻ. Among the Latin authors the vowels that appeared most frequently were the E and also the I. These then are my observations on the number of vowels.

Now I come to the order of the vowels. I noticed the following about the order. A vowel is either followed by another vowel (without there being any consonant between them) or there is a consonant between vowels. First the vowels. After the A, I seem to have noted, it is not uncommon to find in the same word a U, sometimes the O as well, and forming a diphthong, an E; as far as the I is concerned, the Latin authors did not often use it after the A, although the ancient poets wrote 'musai', 'animai' and so forth. Instead, after the vowel O is sometimes found the I and also the U, but this is rather rare, while more frequently the O is followed by the U used as a consonant, as in 'ovem'; frequently the letter O itself is followed by E. In truth, in the Latin authors I have not found A placed after O. With regard to the E, it is associated with another E, as well with all the other vowels in equal measure, each of which individually properly accompanies that letter. In

---

[2] Translator's note: In the Latin, *numeri rationibus*, correctly translated as 'rapporti numerici', that is, 'numeric ratios' in the 1994 Italian version, but as 'numeric relation' in the 1998 English translation, which slights the mathematical intent.

[3] Translator's note: Albert uses the ratio of *octava*, octave or 2:1, translated in the 1994 Italian version as 'mezzo', one half, which misses the point of Alberti's ratios and mistranslated altogether in the 1998 English translation as 'an eighth'. Many thanks to Dr. Monica Ugaglia for pointing this out.

[4] Translator's note: Alberti's ratio is a *sesquitertiam*, or 4:3. The phrase *non excedere consonantes ferme ex proportione quam sesquitertiam nuncupant* is translated in the 1994 Italian 'le consonanti non eccedono generalmente [i vocali] per più di un terzo', and identically in the 1998 English version 'the consonants do not generally exceed the vowels by more than one third'.

[5] Translator's note: Alberti specifies a consonant V and vocalic V, and consonant I and vocalic I. The letters reproduced here are exactly those used in Buonafalce's Latin critical edition.

like fashion the U and the I appear in consecutive geminates as in 'suus'. They can be followed by all of the other vowels.

Regarding the order of vowels, the following considerations must be made: any vowel whatsoever can be found without distinction in the first or last or position or in the middle of a word, as in the word 'aura', but this is not so for the consonants. Finally, we should not overlook which syllabic combinations are created between consonants and vowels and, in the same way between consonants and consonants, which vowels are associated within a single syllable or in the successive syllable, because this is not possible between all of them, as you shall see.

V

Up to this point we have discussed the vowels. Now it is time to speak of the consonants, first their number, then their ordering.

I have observed that G occurs very rarely in writing, F also and so too B must be classified among the more rarely used, followed right after by the letters C, L, Q and K. The consonants I have come across most frequently are S, T and R, whose frequency almost exceeds that of the vowel O, and are outnumbered only slightly by the vowel A. Next to these three in frequency are found M and N.

Now I come to the order of the consonants in a word. Consonants are ordered either by combination or by sequence. Both combination and sequence are of interest. We speak of combination when one or more consonants integrated with a vowel form an entire syllable, as in 'stat'. We have sequence where a letter, either a consonant or a vowel, is disjoined from the one that precedes it, and this disjunction makes it so that each of them makes an independent syllable, such as 'ar-ma', 'cor-pus'. Thus we will first discuss the combination of consonants and then sequence.

VI

Consonants and vowels can be combined in various ways to make a syllable. Since in any given syllable there is by necessity at least one vowel, it follows that either the syllable in question is constituted solely by a vowel as in 'a-er', or that the vowel is combined with one or more consonants, which either precedes or follows it. In some instances a single consonant appears before or after the vowel; in some instances the vowel precedes more than one consonant; in some instances more than one consonant precede the vowel; in some instances the vowel is situated in an intermediate position with more than one consonant coming before and after it.

There is also the case of syllables constituted of a vowel before which there can be combined with up to three consonants that precede it, as in 'scribo', while after the vowel are found at most two consonants, as in 'stans'.

This fact should be remarked as well: between two vowels in any Latin word whatsoever there are never more than four consonants, regardless of the position within the word, and, of these four, the first after the vowel is hardly ever other than the letter B and also D, and sometimes the N, as in 'adstrictus', 'subscriptus', 'transtra'. Any vowel can combine naturally any individual consonant, except the letter Q. Indeed, Q is always followed by a U. In regard to this it amazes me that in writing nowadays the Greek letter K has fallen out of use, the lack of which we notice in the transcription of various terms of Greek origin such as 'kelym', 'Kalendas' and so forth.

On the other hand, when it was established that the U is combined with the Q, it doesn't appear to have been taken into account that the U itself is implicit in this letter Q, so that it sounds like KU. In my opinion the letter used to write 'cespitem' and 'Ciceronem', is not the same as that found in 'consulem', 'curiam', 'causamque', and so forth. But we shall discuss this elsewhere.

Further, X only rarely precedes a vowel, except in rare Greek words, as is also the case with the Z, which we sometimes see in Latin words preceding the first vowel; this, however, is not frequent. Up to now we have discussed the way in which single consonants come before the vowels in the first syllable of the word. Now instead we will investigate how a single consonant can be found after the vowel, again in the first syllable. With regard to this, I have observed that, in the first syllable almost all of the consonants can come after the vowels, with few exceptions.

In fact, neither I nor V used as consonants ever follow the vowels in a syllable, and neither does Q.

Moreover, the letter T as well is only rarely found after a vowel. The letter X rarely comes after any vowel in Latin except after the E. Likewise, the letter G is not associated with a vowel that precedes it, unless it is followed by M or N.

Instead, the other consonants (as I have said) freely associate in the first syllable with the vowel before it, without the necessity of its being followed by a given consonant. However, among all the consonants, C, F, and also P, come after the vowel only if geminate. Up to now we have discussed the case in which a simple consonant is added after the vowel in the first syllable. It remains to speak now about a similar situation that is found in the last syllable.

The addition of consonants after vowels in the last syllable of a word depends on whether the word is monosyllabic or polysyllabic. In monosyllabic words the vowels can be followed by all of the consonants except F, G, P, Q, I and V. In polysyllabic words, in addition to the consonants just mentioned, the vowels will not be followed by B, C, D, and in Latin words, L. In contrast, in both polysyllabic as well as in monosyllabic words M, N, R, S, T and X can easily found after the vowels. I need not refer to the letter Z, except for one case, because it is most rare in the Latin authors.

VII

I now come to the consonants that are either associated in groups of two or three. There are various instances. Some of the associations are never found following a vowel at the end of the word in the Latin authors, while other are never found at the beginning of a word, and some instead are found in positions between vowels as well as at the beginning or the end of the word. In pairs of consonants before vowels, especially at the beginning of a word, one will always be either S, L or R.

These are governed in different ways. The letter S always comes first in a pair. Instead, L and R appear as the second in order, and there are fifteen kinds of pairs which are never situated at the end of the word in Latin words, and in seven of these the second letter is R, that is: BR, CR, DR, FR, GR, PR, TR. Instead there are five where the second letter is the L: BL, CL, FL, GL, PL; those in which the S appears first are three, that is, SC, SP, ST, as well as SQ. On no occasion are such kind found at the end. In contrast the consonant pairs that are quite often found at the end of the word and never at the beginning are five: NC, NS, NT, NX and PS. Additionally there are LX and RX, as in 'calx' and 'arx'.

The Tuscan language sometimes puts the S at the beginning of the word before all of the consonants except the letter X, and then there is a consonant pair that is found in all positions, and that is ST, as in 'stat', 'adest', 'restat'. With regard to the final consonants we must not neglect to make the following observations. Of two consonants placed after the vowel at the end of a word, the final one of the two must be: T, S, X, and C as well. These four letters will be accompanied further by penultimate consonants which must be either: B, L, N, R, P and S, with precisely determined variants; of the four, only S will be found after B and after P, after L only T and X will be found. Finally, the letter N can be freely followed by any of the four letters I mentioned above: T, S, X, C.

We have finished with the topic of consonant pairs, except for considerations about which associations of two consonants are used more or less frequently. But we have said enough about this. In the Latin authors the groups of three consonants that are placed before the vowels are SCR, STR, SPL, and also SCL, as in 'sclavus', in addition to these the Tuscan language also uses SBR, SDR, SFR, SGR, SPR, groups that occur in spoken Tuscan without R as well. In all these groups of consonants (as you can see) the first of the three is always an S, the combination of three consonants is also found, although rarely, after the vowel and exclusively at the end of the word, as in 'stirps' and 'urbs'. This concludes the discussion of the combination of consonants before and after the vowel.

VIII

Now I come to the relationships of consonants in sequence. We speak of sequence (as we said) when two consonants are found to be situated in a single word in a position between vowels, such that the first of them unites and forms a syllable with the vowel that precedes it and the other with that which follows. It is certainly clear that, as far as consonants are concerned, like follows like, as in the case of geminations of the kind BB, CC, DD, FF, and so forth, such geminations occur with almost all the consonants, except for X and Q. These are never geminate in words.

Here we shall consider the particular behaviours of some of the consonants taken individually. In a single word, B can be followed in the next syllable by all of the consonants except Q and X and Z. Likewise, the letter L can be followed in the next syllable by all of the consonants except I, Q and R; R by all of the others, including the X, although this is very rare, just as it is also very rare to find the letter F and the consonant I. The letter D can be followed by the consonant I, L, M, N, S, V as well as R and T, though these generally end up attracting to themselves the D that comes before, so that the D is combined to them. Otherwise these letters I have reviewed do not unwillingly follow <the D>[6], the others do not follow without difficulty.

In contrast, within a word, there are some consonants that never appear in the next syllable after V and the consonant I, while following C there is only the C and also the T, and following F we only find the F and, in Greek, also T. Following the letter P there are only T and S, and nothing after T except Q, after the letter G only M and N will follow

---

[6] Translator's note: Insertions in angle brackets < > are the translator's, for clarity.

disjoined in the next syllable, and after the letter M only B and only P, and in some cases also F.

These considerations show that there may or may not be consonant sequences in separate syllables.

In regard it must be observed that the letter D, in a position between vowels, followed by another consonant, is aways preceded by A, as in 'admissus' and 'adiuvo'. That is all that needs to be said of sequences of two consonants.

If instead of two there are three consonants situated between vowels, it will be found that almost everywhere the first of these will aggregate to the vowel that precedes it, while the remaining two consonants will associate with the vowel that follows, as in 'impleo' and 'instruoque'; if it were said that in cases such as 'transportavi' the sequence NS seems to associate itself to the vowel A before it and in the word 'pistrix' the three consonants form a syllable with the final vowel I, I would be reluctant to accept these objections; but cases of this kind, are rare as well as difficult to identify and collect. We note the phenomena that occurs most frequently. In fact, there are practically no Latin words in which three consonants are situated between two vowels in which the first is not either B or D or X or M.

So much for groups of three consonants. The case of sequences of four consonants situated between two vowels is similar: one first consonant will form a syllable with the vowel that comes before it, the remaining three with the one that comes after, except perhaps in this case: 'transfretari', where the first two consonants are associated with the vowel before them. It can be affirmed: wherever there are four consonants between two vowels, the last three will always be like those what we mentioned above, that is, those that come before the vowel in a three-consonant combination. It is also worthy of note that of four consonants situated between two vowels, the first is almost always only B or D or N, while the last instead is L or frequently R.

IX

We have discussed the number and the order of the letters and those vowels that are most frequently encountered in writing, of the possible association of vowels with regards to the consonants, we have said which are most frequently encountered and regarding their arrangement, how they behave in addition or in sequence. From what has been said thus far you will quite easily deduce (if I am not mistaken), how astute minds will perceive a way to decipher and interpret coded writing in a most easy way.[7] Once the signs and various graphic characters present in the missive have been extracted and gathered, in proportion to their frequency, it can be conjectured that the null characters, as they are called, or the duplicate ones, that is, those corresponding to the same letter, are as many as there are in excess over twenty; since in writing we use only nineteen signs. These are: A, B, C, D, E, F, G, I, L, M, N, O, P, Q, R, S, T, V, X; with the addition of Z, if you like, which makes twenty. All of the other ciphers are either superfluous or duplicates.

Among them all there will be vowels which occur more frequently and are not very far apart, and the consonants can be identified by their greater or lesser frequency on the basis

---

[7] Translator's note: Thanks to Dr. Monica Ugaglia for clarifying the proper word order in this sentence.

of the observations reviewed above. Moreover, through consideration of the combination and ordering, little by little the hidden scheme of how they have been written will be revealed.[8] To this end, however, it is above all careful investigation and determination that matter the most. There is in any case one observation that has not yet been made and which I think is worth making, which is that I have observed that a vowel, within a single word, can be followed by another without their being any consonant between them. Most of the time a single consonant is inserted between two vowels in the same word, fairly often two, rarely three, and even rarer still, four.

X

The letters, their intrinsic qualities in relation to their common use in the Latin language have been discussed. Now we shall tell of the various kinds of devices that are used in the invention of coded languages. First, however, it is best to explain a little more clearly which expedients are useful for making an encrypted text more difficult to decipher. Care must be taken that the characters that express the most frequently occurring vowels and consonants in the writing are overabundant and dissimilar; there should thus be several marks denoting E and likewise R and so on, in writing we will not always use the same character, but of those at our disposal, we will use first one and then another.

It will be to our advantage to ignore the rules of spelling. For example, writing 'arrogans' with a single R, and in the same way never using geminates in any word. We won't add the letter U to Q, nor will we ever put an H; we will establish different signs for the consonant and vocalic V. It is most useful at the beginning of the missive, but elsewhere as well, to write words in which the vowels or consonants have been taken away, or which are meaningless. These observations seem not without usefulness for those writing in code.

Now I return to the matter. Taking up the discussion started at the beginning, a cipher is a system of writing whose signs take on an arbitrary meaning agreed upon by the writers. It is therefore necessary to consider the characters used and the conventions adopted by the writers.

XI

The first point. Note that some of the letters are those used by the Latins such as A, B, C, and so forth; others are not generally in everyday use.[9] Among those not commonly used we have not only those used by the Greeks or Arabs or other peoples in their written texts, but above all those created by ingenious minds, constructed for example from points, figures and other new inventions.[10] The meaning assigned to them will actually be arbitrary, not based on any resemblance between sign and nature, as is the case of the characters carved on the ancient Egyptian obelisks, but rather determined by the choice of convention by the person who created the writing. However determined, the individual signs can express the meaning of simple letters or syllables or words, or can even indicate entire sentences.

---

[8] Translator's note: Thanks to Dr. Tessa Morrison for clarifying this sentence.
[9] Translator's note: Thanks to Dr. Tessa Morrison for clarifying this sentence.
[10] Translator's note: Thanks to Dr. Tessa Morrison for clarifying this sentence.

Thus a common letter, say A, will take on the meaning of another letter, say G, and the letter B will indicate another letter, say, M, and in this way letters are used to obtain new unusual meanings different from the traditional ones; or, in their place, we will use alternate signs which don't have anything to do with the characters habitually used. This is the only method of monoalphabetic encryption which uses simple letters. In addition to what was said above, it is possible to make more characters both usual and unusual, and to use them individually as well as in combinations, whether they are two, three or more, to represent the value of a letter.[11]

Conversely, a single sign can indicate several letters. This especially regards those that we call combinational, because, in writing, they are very often found combined, as is the case with the majority of the consonants that follow the letter S or come before the letter L and the letter R. It can also be established that the alphabetic characters or graphic signs signify a syllable or even a word or an entire phrase, as when A means 'pontiff' and B 'army', D 'fleet' and in the same way R could mean, for instance, 'the enemy moved camp', S 'the troops have no provisions', and so on, as you like. All of this can be decided arbitrarily, utilising one, two or more signs, be they common, unusual or obsolete.

Further, it is possible in writing the order of the individual letters can be scrambled; imagine that the first letter of a word is placed last in line, the second letter of the same word next to last, the fourth in place of the second and so on.

Once a certain rule is established for the distribution of the letters, the rest follows in consequence, so that it is understood how it is likewise possible to transfer the characters from the first to the second or any other line of the text; the relations are practically infinite, and given the variants, it will be highly encrypted, although the order of the consonants will become explicable if a shrewd and acute expert in encryption put his mind to it.

All of what we have said about switching the letters and on confounding and inverting the logical order is valid for the syllables as well. Moreover, an entire word can take on the meaning of another word, as in the case where the preposition 'pro' signifies 'ad', 'in' signifies 'sub' and so forth, nouns express the meaning of other nouns, as if by 'book' we mean 'fleet', with 'field', 'legion', or they might be transformed into pronouns, as if 'I' signified 'pope', 'you', 'consul', et cetera. Verbs can also become nouns, by making correspond the noun endings correspond to vocalic actions and tenses, such as in the case of 'father' for 'read', 'of father' for 'reads', 'to father' for 'has read', 'by father' for 'reading', and similar devises that aren't worth going into.

Further, it is also possible to insert into the text in practically endless ways not only syllables but especially words; of these systems one of the most effective consists in taking a book of poetry or prose, or composing a fake letter to <someone> who is intimate, into which apt words and expressions are scattered in a suitable position. So distributed they must be identified and collected by the friend far away, these words have to be evidenced by means of some precise signs agreed upon by convention.

As signs, marks such as the period, comma, an inverted comma between the lines or in the margins, or an erasure or scratching out here and there won't raise the suspicions of an investigator. In order for the mark to be less evident to the eye of a skilful code breaker it

---

[11] Translator's note: Thanks to Dr. Tessa Morrison for clarifying this sentence.

could be established that it is not the words thus indicated that are to be collected but those that precede or follow it, or is opposite to it, or that found a certain number of words or lines away from the agreed-upon sign.

What has been said about the letters, syllables and words is also effective for the complete phrase as well, making them take on a different meaning and scrambling the structure so that to an outsider the words appear like the leaves of a tree blown about by the wind that have been raked into a pile and set there.

XII

There are in any case other similar systems of writing that may perhaps be pertinent. But these that we have discussed so far will suffice. Perhaps it will please you to know about ineffective expedients that are considered by some to be valid, such as writing with milk, acid solutions, onion juice, and other such eccentricities, so that the message can't be read unless it is heated over a flame, sprinkled and smeared with dust, or dampened with water treated in some way, or exposed to the sun's rays. These are useful. Then, of the expedients of which the ancients tell, of the arrow, the scytale, the hare, the shaven man, not only don't they merit preference them over better methods, but don't even merit putting any faith at all in them. However, I will discuss, for your pleasure, one of our most secret devises.

There are some parts in the human body even more concealed than a horse's hoof, where it is possible to write a text of some length with a liquid such that, even after a period of more than twenty days it is still quite readable, and which in the meantime isn't erased by sweat, water, or heat. After a compress of a special medicated water is applied to the parts where the message has been written, they contract and wrinkle up into a ball so rough and dense that no trace of the writing can be discerned. We can then make it open up again with another liquid for easy reading. But otherwise it remains hidden.[v]

XIII

Now in what follows we will tell about the writing system of our own invention. It is very commodious, no other cipher is more expedient, easier to use, faster to write with, none quicker and more rapid to read, none (the index agreed upon between me and the other to whom I am writing being unknown) more difficult to decipher. I can affirm that the most acute and diligent intellects, the constant application of the most perspicacious minds, every speculative skill and all effort, will be frustrated. Not even the most diligent, unless they know how it is constituted, can understand my taxing cipher.

I also add that any common letter-carrier who is called to take dictation, even though the letters are usual and known, will remain ignorant of what he has written, likewise a person who reads a missive that has reached you from far away; while you can completely understand the contents, that person, to whom you have given the letter to read, will not be able to understand a single syllable, thus I can justly consider this cipher worthy of sovereigns, who can use it quite easily, with little effort and without being encumbered by use of an interpreter.

Enough said. Let's get on with it. This is the system I consider to be the most cryptic and commodious: make two circles cut from two bronze sheets, one larger, called fixed, the other smaller, which we call mobile. The diameter of the fixed circle is larger by the ninth part.[12] I divide the entire circumference into twenty-four coequal parts; these parts of the circle are called houses. In the individual houses of the larger circle I write in red ocher upper-case letters following the usual alphabetic order, that is, first A, second B, third C, and so on with the rest. The H and the K are left out because they are used so rarely. These upper-case letters will be twenty in number, as we said above, and will occupy twenty houses, which will be called the houses of fixed and real letters.

Those four empty houses that remain will be called numerals because of the fact that in each of them will be written a number in small characters, with black ink, that is, in the first 1, in the second 2, in the third 3 and in the fourth 4. Thus, in this way, all of the houses of the large disk are filled.

In the smaller circle there will then be houses similar in number to those of the larger disk and with corresponding lines dividing them into houses, which will be called mobile, will be written a letter in black ink, not in upper-case letters, but in lower-case, and not in alphabetical order as in the fixed circle, but strewn by chance, say that the first of these mobile letters is a, the second g, the third q, and so forth with the rest, until all twenty-four houses are filled. There are that many Latin alphabetic characters, the last of which is &.

This done, we place the mobile circle on top of the larger fixed circle such that a single pin pierces the centre of both acting as the axis, so that the mobile circle rotates.

The instrument composed of the two circles we call 'formula'. It is indispensable to have two exemplars of the formula: one for you to keep and the other for the friend far away to whom you will write, which two formulae will be alike in terms of position, number and order of the letters, and in no way discrepant.

This done, we must establish what the index will be; this index is a genuine key that opens the access to the most tightly guarded secrets. The indexes are two; one is an index drawn from the fixed upper-case letters, the other from the mobile lower-case letters, both arbitrarily chosen.[vi]

XIV

First the mobile index.[vii] Say for example we have mutually established k as the index of the mobile circle. Writing, the formulae are positioned at will, say such k lies under the upper-case B and the next letter corresponds to the letter that comes next. In writing to you, I will first of all put the upper-case B under which lies the index k in the formula; this is a signal to you far away, wanting to read what I have written, that you should set up the twin formula in your keeping, positioning the mobile circle so that the B sits over the index k. Then all of the rest of the lower-case letters present in the coded text will take their meaning and sound from those of the fixed circle above them.

After I have written three or four words I will mutate the position of the index in our formula, rotating the disk let's say, so that the index k falls below the upper-case R. Then in the missive I write an upper-case R to indicate that k no longer refers to B, but to R, and the letters that follow will assume new meanings.

You likewise, far away and receiving the message, have to look carefully in reading to find the upper-case letter, which you will know serves solely to indicate the positioning of

---

[12] Translator's note: One and one ninth is 10 : 9, the minor tone. See [Alberti 1988: 305 (IX, v)].

the mobile circle and that the index has changed. Thus you too will position the index under that upper-case letter, and be able to read and understand the entire text with ease.

The four mobile letters that under the four houses on the fixed circle above that are marked with numbers, regardless of the values they themselves have, do not (if you will) receive any meaning, and can be inserted into the text as null letters,. However, when combined or repeated, they are marvellously commodious, which I will describe below.

XV

Alternately, an index could be selected from among the upper-case letters and we could mutually agree on which would be the index; say we have determined the letter B as the index.[viii] The first letter that will appear in the missive that I write to you will be whatever lower-case letter you want, say q; the *formula* will be positioned so that this lies under the index B. It follows that q will take on the phonetic and semantic value of the B. Finally we shall follow all the rest of the writing, as we said, with regard to the index of the highest mobile <circle>. When it is then necessary to modify the encoded alphabet and the positioning of the formula, then I will insert into the missive, in the proper place, one and only one of the numeral letters, that is, one of the letters of the small circle lying under the numbers that signify, say, the number 3 or 4, and so on.

Rotating the mobile disk, I will make this letter correspond precisely to the agreed-upon index B and, successively, as the logic of writing requires, I will go forward, assigning to the lower-case letters the value of the upper-case letters.

In order to further disorient the investigators, I could also agree with my friend to whom I am writing that the interposing upper-case letters (of which without this convention there would be none) have no value, and similar other devices that it is not worthwhile listing. Thus, by positioning the index in a different way by rotating the mobile disk, it is possible to express the phonetic and semantic value of each of the upper-case letters, using (as you can see) twenty-four different alphabetic characters, while each of the lower-case letters can correspond to any upper-case letter whatsoever and also to the four numbers of the disk above.

I now come to the use of the numeral letters, of which is nothing <more> admirable.

XVI

The numeral letters are, as I said, the lower-case letters that correspond to the four numbers of the fixed circle above. The numbers, combined in groups of two, three or four, three hundred thirty-six whole phrases determined at will.

These numeral letters, when paired, say that ps corresponds to 12 and pf to 13, and with other similar pairings that can for constructed with these four numerals, indicate sixteen phrases. If instead these same numbers are combined into groups of three, say psf signifies 123 and sfp 231, sixty-four phrases can be expressed. With combinations of four numerals, where sfkp is equivalent to 2341 or fpsk corresponds to 3124, and similar combinations, 256 whole phrases can be expressed. The total sum of whole phrases is 336. Now we shall show how these numeral letters are used.

On one side we compose a table of 336 lines, in which we clearly arrange the numeral combinations at the beginning of the line, that is, in the first line there will be 11, in the second 12, in the third, 13, in the four 14, in the fifth 21, in the sixth 22, in the seventh 23, and so forth for all the rest, as in the table we show below. We ascribe to each individual line of the table, next to the corresponding number, say next to the number 12 'the promised ships have been equipped and provided with provisions'.

In a similar way whole phrases with whatever contents we want are ascribed to each number combination in the table.

It is necessary for you to have a copy of this table with you so that when you who are far away receive my letter and you come across the numeral letters, noting that they signify numbers, you will consult the table that contains the predetermined phrases written there; for writing I will say that there is no invention that is quicker, more secure and nothing devised for cyphers could be more aptly suited, that two, three or four letters combined in different ways can express three hundred thirty-six phrases, isn't this wonderful?

It is recommended that I have with me two numeral tables and for you to have two as well, arranged in different orders, one where we set out the numbers at the beginning of the line so that they are easy to read; in the other conversely in alphabetical order will be arranged the phrases according to the letter that forms the title so that the writer can find them quickly. The phrases will be arranged in the following way: those that regard administration of provisions (*annonam*) will appear under the initial A; those that refer to the carrying out of military operations (*bellum administrandum*) under B; those that regard ships (*naves*) under N, and so forth for the rest.

The difference between the two numeral tables is that, in the first, at the beginning of the line, there appear the numbers followed by the phrases, conversely what appears at the beginning of the line is not the numbers but the initials followed by the phrases, and at the end will be the numbers that have been attributed to the phrases in the correlated table.

Thus, in writing to you, I first look up the phrase that I want to use in the table, and having found it under the corresponding initial, I look at the numbers noted and using our encrypted *formula*, I insert the letters that signify those numbers into the missive. You, as I have said, will deduce the phrase from those numbers.[ix]

I would have this little work of mine kept among our friends, not in the public domain, so as to not profane a subject worthy of sovereigns and rather for statesmen devoted to the most important of affairs. Be happy.

XVII
Formula [x]

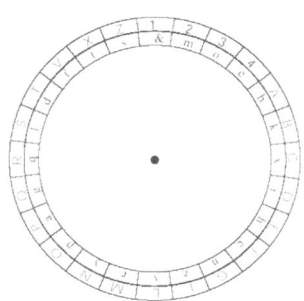

XVIII
Numeral Tables [xi]

Numerical Tables from Ms. Miscellanea Correr 47 2130, fol. 7r. Reproduced by permission, Biblioteca del Museo Correr

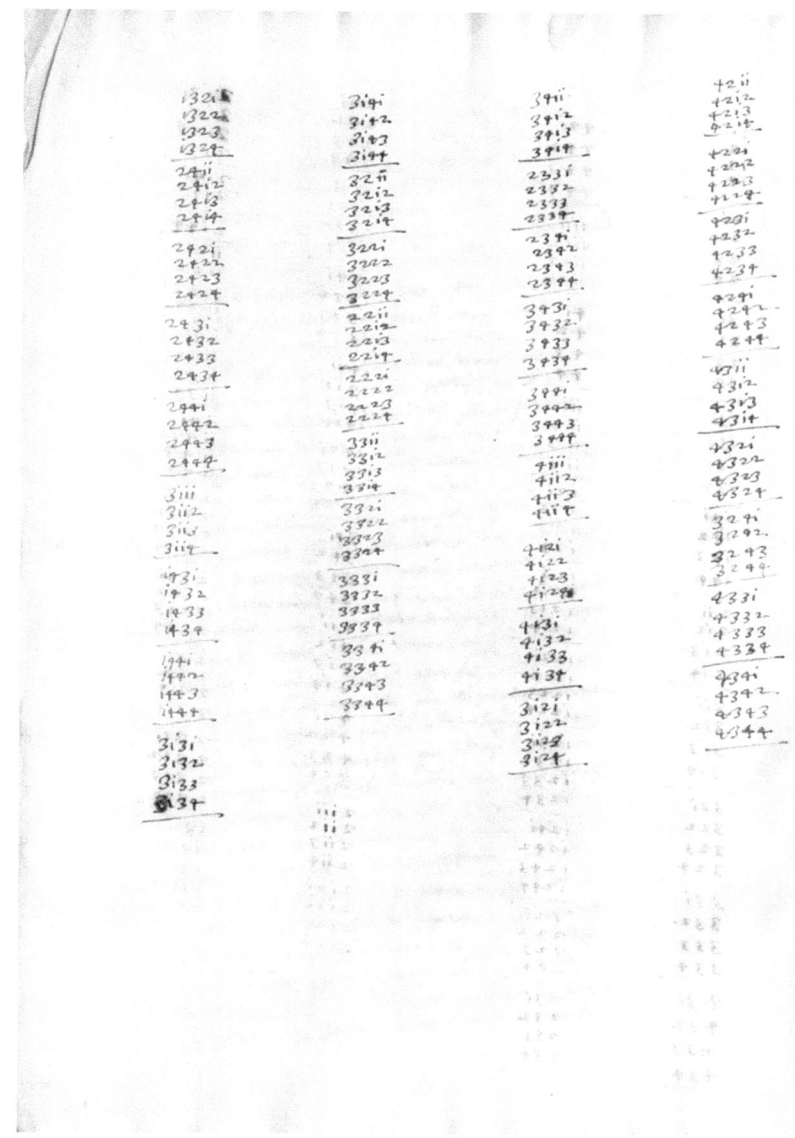

Numerical Tables from Ms. Miscellanea Correr 47 2130, fol. 7v. Reproduced by permission, Biblioteca del Museo Correr

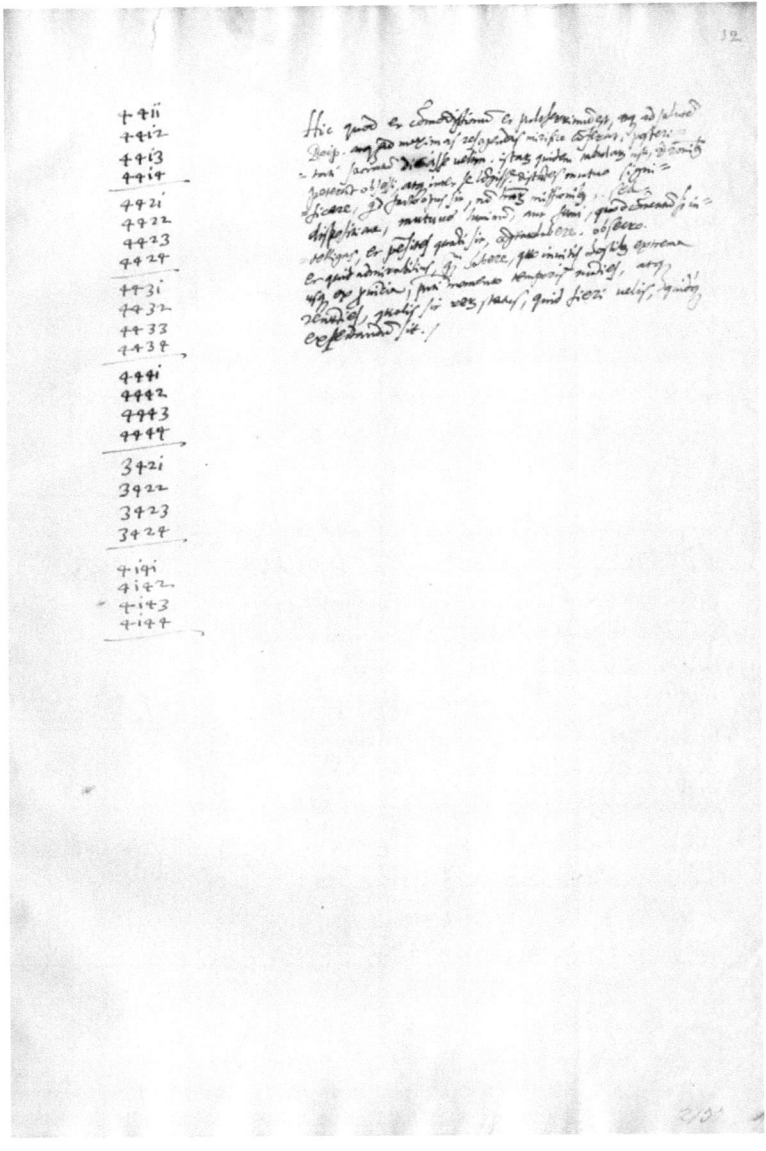

Numerical Tables from Ms. Miscellanea Correr 47 2130, fol. 12r. Reproduced by permission, Biblioteca del Museo Correr

XIX

I would like to consign this beautiful and useful work, which can contribute in an extraordinary way to the preservation of the Republic and to the treatment of political affairs, to posterity. Through the shrewd use of these tables, it will be possible for those under siege and separated by long distances to communicate what should be done, not by sending missives, but by means of signals made with lights or smoke. When you think about and understand this, you'll congratulate me.

I beg you[13]: for what would be more marvellous than to have the means by which, against the wishes of your enemies, you could send messages and cancel others in the briefest instants of time even from the most far province about the state of play, about what you would wish to happen and what you expect to happen? Farewell.[xii]

---

[13] The first word in Latin is *obsecro* , 'I beg you' but one has the impression that something to which 'obsecro' refers has dropped out. My thanks to Tessa Morrison and Richard Schofield for help with this sentence.

Lionel March

Chapter I

i. Alberti points out the two aspects of what is today known as crytptology, the science of making and breaking coded messages. Cryptography is the science of making secure codes, while cryptanalysis is the science of breaking them. In the eighteen short chapters of this paper, Alberti first addresses cryptanalysis (Chapters IV – IX), and then, following an intermission, makes an original contribution to cryptography (Chapters XIII – XVIII).

Chapter II

ii. Alberti is in the Vatican gardens in Rome with his longstanding friend and humanist Leonardo Dati (1408–1472). Dati was then chief secretary to Pope Paul II. In the 1440's Alberti and Dati were involved in a literary competition which was designed to match the quality of classical Latin poetry against verse in contemporary Tuscan. Dati submitted a winning entry on the theme of 'friendship', while Alberti took the opportunity of producing the first grammar of the Tuscan language. Over the years, Alberti had often submitted his writings to Dati for editorial scrutiny, starting with the martyrology of Saint Potius in the early 1430s, at the same time Alberti joined the papal chancery as an *abbreviatore apostolico*.

Most likely the German inventor was Johannes Gutenberg. However, Dati and Alberti were familiar with the partnership of Arnold Pannartz and Konrad Sweinheim which came to Rome in 1467 from its first monastic workshop founded three years earlier under the patronage of Paul II. The partners are celebrated for their introduction of Roman typefaces replacing the Gothic used by Gutenberg and others. Recall that Alberti had strong interests in architectural typography.

At some time, during the brief reign of Pope Paul II (1464-1471), young Luca Pacioli stayed in Rome as Alberti's guest, as Pacioli later records in *De divina proportione*, where rationalized Roman capitals are shown. Alberti was well acquainted with Pannartz and Sweinheim's editorial advisor and proof-reader, Giovanni Andrea Bussi, Bishop of Aleria, to whom Alberti had only recently dedicated his manuscript *De statua*. Anthony Grafton suspects that Alberti was hoping, through this dedication, to have his short manuscript printed.[1]

---

[1] Cf. [Grafton 2000: 331–332; 336].

Alberti ceased to be an *abbreviatore apostolico* in 1464. Two years later, Paul II annulled the College of Abbreviators altogether, amid protests from humanists who were suspected of entertaining ancient Roman, pagan sympathies. Humanist poets and men of literature – including Alberti – had benefitted from sinecures as members of the college under previous pontificates. A number of sacked abbreviators were imprisoned and tortured for conspiracy against Pope Paul II. Among these disillusioned and persecuted humanists would have numbered close acquaintances and former colleagues of Alberti's. Yet Alberti's fellow colloquist, and lifetime humanist friend, had himself risen to be chief secretary to this very same Pope.

iii. It should be recalled that Pope Pius II, under whom Alberti served as *abbreviatore apostolico*, had initiated a crusade against the Turks in 1460 which effectively ended with the Pope's death in 1464. It is in such circumstances that secrecy is required in communications between the Vatican and its allies in the field and vice-versa. Yet after 1464, Alberti would have been acutely aware of the situation his humanist colleagues found themselves in vis-à-vis papal informants seeking evidence of pagan and disloyal behavior. In particular, the needs of secret communications between members of the Accademia Romana, which was dissolved in 1467 by Paul II, with whom Alberti had some sympathies. In both cases, public and private, encoded correspondence was an essential mode of communication.

Fig. 1. Alberti's façade for Santa Maria Novella. From [Tavernor 1998: 105]

Dati recalls that Alberti 'has always been involved in investigating occult arts'. A contemporary work that was nearing completion in the 1460s was the front elevation of Santa Maria Novella (fig. 1), where the Florentine's familial and namesake Leonardo Dati (1360-1425) had been Prior from 1401; and who, around 1424, had communicated with the twenty-year old Alberti concerning the latter's *Philodoxeus*.[2] In the façade, Alberti displayed both mathematical skills in the proportional system and esoteric knowledge through the numbers [March 1997, 1999].

The façade is in four levels. On the first level the side pilasters are striped alternately in equal green and white marble bands. There are 15 green bands, and 14 white. On the next level the side pilasters have 6 green bands, and 5 white. On the third story the pilasters have 13 green bands, and 12 white. These numbers all relate to known rational values for √3. The ratio 26:15 is a good rational proxy for √3, and consequently the ratio 15:13 (of lower green stripes to upper) is a proxy for √4 : √3. Similarly, with the white stripes, the ratio 14 : 12 = 7 : 6 is an earlier proxy for √4 : √3 in the process of deriving improved rational values using 12 : 7 for √3. The ratio 6:5 – green to white – in the middle section is again √4 : √3 using the value 5:3 for √3. This becomes clear in the classical rational derivation for √3. Let p:q be a rational value of square root 3, then 3q : p is also a rational value, since $p/q \times 3q/p = 3$. The ratio $(p+3q):(q+p)$ is a better rational value:

p : q	3q : p	(p + 3q) : (q + p)
5 : 3	9 : 5	14 : 8 = 7 : 4
7 : 4	**12 : 7**	19 : 11
19 : 11	33 : 19	52 : 30 = 26 : 15
26 : 15	45 : 26	71 : 41

If there is any doubt about Alberti's pervasive mathematical skill, note that the two giant scrolls screening the aisles roofs are based on two circles (fig. 2). The outer is divided into 26 radii; and the intermediate into 30 radii – a ratio of $30:26 = 15:13$, again the rational proxy to √4 : √3. In the centre of each scroll is a polygonal design; one an octagon and the other a hexagon – a ratio of $8:6 = 4:3$, sesquitertia. Alberti is giving reign to his proportioning concept of *correspondentiae innatae* (radices and potentiae, roots and powers) derived from geometric forms, in this case the equilateral triangle, the symbol of the Trinity. If √4 = 2 is the side of the equilateral triangle, then √3 will be the altitude. The square on the side is 4, and the square on the altitude is 3.

---

[2] [Grayson 1954: 291-293].

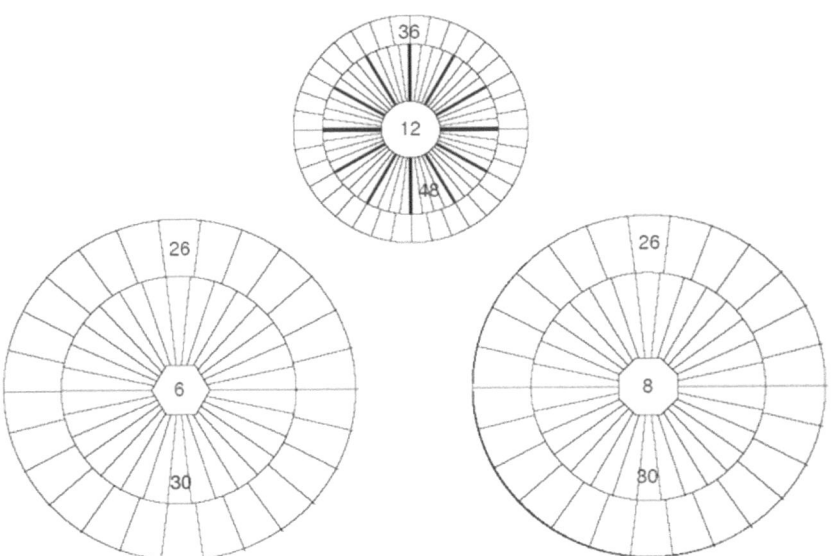

Fig. 2. Radial divisions of the circles in the scrolls and tympanum on the façade of Santa Maria Novella

The fourth level is the tympanum. In the centre of the tympanum is a circle, divided this time into 36 outer radii. The inner circle is divided into 48 radii, the same as the circular disk known as the *horizon* used by Alberti in his survey of Rome. Thus the ratio of inner to outer radii is once again 48 : 36 = 4 : 3. The greatest common factor of 36 and 48 is 12.

The circle contains, it is said, the crest of St. Bernardino of Siena which iconographically carried the letter IHS for Jesus in its centre. In this case, the monogram is not present and the sun radiates around a face giving emphasis to 12 rays out of 48. In the centre is the face of Sol, the sun, or the zodiacal Leo. Alberti's signature is all over the façade starting with Leo. Take the intermediate radii from top, left and right: 12, 30, 30. These numbers 'spell' Albert's full Latin name.

Using the single digit nine square Latin encoding where, for example, O(50) and Z(500) both go to 5 (fig. 3):

A(1)	B(2)	C(3)		K(10)	L(20)	M(30)		T(100)	V(200)	X(300)
D(4)	E(5)	F(6)		N(40)	O(50)	P(60)		Y(400)	Z(500)	
G(7)	H(8)	I(9)		Q(70)	R(80)	S(90)				

Fig. 3.

LEO	$(2 + 5 + 5)$	$= 12$
BAPTISTA	$(2 + 1 + 6 + 1 + 9 + 9 + 1 + 1)$	$= 30$
ALBERTVS	$(1 + 2 + 2 + 5 + 8 + 1 + 2 + 9)$	$= 30$

When reduced to single digits, (12): 1+2=3, (30): 3+0=3, (30): 3+0=3, Alberti's name becomes 3+3+3, a Trinity of Trinities and the sides of an equilateral triangle, the regular polygon from which Alberti's *correspondentiae innatae* are derived for the façade of Santa Maria Novella. Using the full digit coding:

LEO	$(20 + 5 + 50)$	$= 75$
BAPTISTA	$(2 + 1 + 60 + 100 + 9 + 90 + 100 + 1)$	$= 363$
ALBERTVS	$(1 + 20 + 2 + 5 + 80 + 100 + 200 + 90)$	$= 498$

the whole name sums to $75 + 363 + 498 = 936$, the product of 26 and 36, exactly the outer radii of the three circles.

In Hebrew gematria similar to the Latin nine square above, but known at the time through Christian cabbala, the name of God – YHVY, Yahweh – sums to $26 = \text{Yod}(10) + \text{Hay}(5) + \text{Vav}(6) + \text{Hay}(5)$. For example, it is surely no accident that Bonaventure (1221–1274) counts 26 books of the then current Old Testament as being divided into 5 legal books, 6 historical, 5 sapiential, and 10 prophetic.[3] The Tetragrammaton displays YHVH in a triangular formation:

YHVH	$10 + 5 + 6 + 5$	$= 26$	$1 + 5 + 6 + 5$	$= 17$
YHV	$10 + 5 + 6$	$= 21$	$1 + 5 + 6$	$= 12$
YH	$10 + 5$	$= 15$	$1 + 5$	$= 6$
Y	10	$= 10$	1	$= 1$

where $26 + 21 + 15 + 10 = 72$. 'Leo' was a name adopted by Alberti in mid-life. It added 12 to 60, the sum $30 + 30$ of his given names. This made his new full name in Latin equal to 72, a name – LEO BAPTISTA ALBERTVS – that was surely divinely protected in God's name. In single digits when 10 is reduced to 1, the Tetragrammaton sums to $17 + 12 + 6 + 1 = 36$ and still heralds Gods's name. The outer radii – 36, 26 – are thus seen to be impregnated with divine significance. Are these recondite relationships intentional or merely coincidental? They are within the realms of renaissance thought, where mathematics often had an occult, even magical status. Especially is this so where certain ideas had permeated to Rome from Spain. There, Christian cabbalists were using arithmogrammatic 'calculations' to demonstrate theological, particularly christological truths. It is this commentator's conviction that Alberti and Dati could not have been unaware of the Vatican's surveillance of these esoteric matters

Chapter III

iv. Alberti sets out his approach, first to make an analysis of the plain text mostly in Latin, but occasionally in Tuscan, and second to propose an original coding system to produce a cipher text. He makes no acknowledgement of predecessors. Recent discoveries have revealed that al-Kindi, the philosopher of the Arabs, had made a statistical analysis of

---

[3] See [Monti 2005].

Arabic plain texts, circa 850 CE, some 600 years before Alberti's contribution for plain texts in Latin. Alberti's work may have been helped by his previous experience in producing the first dictionary of Tuscan grammar.

Chapters IV to XI are self-explanatory and demonstrate Alberti's meticulous enumerations of the frequency of vowels, consonants and syllables.

Chapter XII

v. Alberti is sketching a brief account of steganography, the art of concealed writing. He refers to an instance recorded by Herodotus: a man had his head shaved, the message was tattooed on his bald head, and then the hair was regrown to conceal the writing. Alberti also refers to the Spartan *scytale* which comprised a leather strap. This was wrapped around a staff of a certain diameter, and the message was inscribed, left to right along the staff, one letter to each wrapped strand. The message appeared scrambled when the strap was unwound. It required a staff of identical diameter for the message to become apparent when the leather strap was rewound.

The last method is described by David Kahn, who avoids the obvious question as to what part of the body Alberti refers: 'some parts in the human body even more concealed than a horse's hoof'. Kahn describes how the Chinese would write on very thin paper or silk which was then rolled into a ball and waxed. The waxed ball would then be swallowed or hidden on the messenger's body, possibly in the rectum.

Chapter XIII

vi. Alberti describes his invention of the formula. There has been some speculation that the circular design might have been suggested by an instrument in Ramon Lull's *Ars Magna*. Lull (1232–1315), a Majorcan, was a secular Franciscan who is often associated with the beginnings of Christian caballa. In the two centuries following his death his works were condemned twice by the Vatican. In Lull's wheel nine concepts are rotated against one another to suggest new philosophical/theological concepts (fig. 4). In turn, Lull's instrument may have derived from the Arab astrologers' *zairja*.

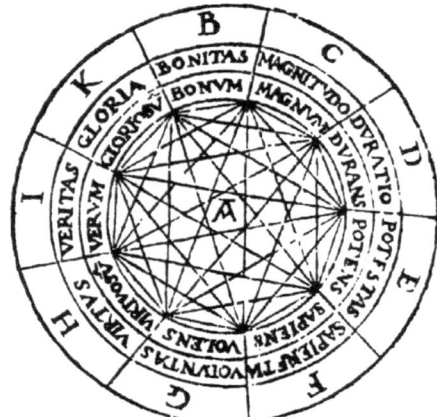

Fig. 4. Ramon Lull's circular instrument from *Ars Magna*

A closer source of inspiration must have been Alberti's own interest in the relationships between outer and inner circles shown in his contemporary façade design for Santa Maria Novella. That one circle in the *formula* rotates, as do circles in Lull's wheel or the astrologers' *zairja*, is not something that requires a great leap of the imagination.

The division of the circles of the formula by 24 radii is required by the 23 letters of the Latin alphabet: A B C D E F G H I K L M N O P Q R S T V X Y Z. However, 24 is also aesthetically consistent with the geometries of Santa Maria Novella where 48, 36, 12 radii are presented. Alberti only uses 20 Latin capitals, omitting the aspirant H, and the letters K and Y whose infrequent appearance in plaintext had been noted above in his analysis of Latin's usage. Alberti marks these 20 uppercase letters in red and places them in alphabetic order around the outer circle. He fills the remaining four 'houses' with the numerals 1, 2, 3, 4 in black. This completes the stationary part of the *formula*, or *stabilis*. When it comes to the inner circle, the *mobilis*, Alberti uses all 23 lowercase letters with the addition of the ampersand, &. These are sprinkled at random around the circumference of the *mobilis*. An exact copy of the *formula* is forwarded to the friend to whom coded missives will be sent. Since Alberti and Dati started their conversation concerning printing practices it is interesting to note that while Pannartz and Sweinheims' Roman typeface contained the expected 23 capitals, their font contained three times as many lowercase graphemes, including diacriticals and ligatures.

Chapter XIII

vii. Figures 5 and 6 can clarify Alberti's explanation of the use of his *formula*:

> Say for example we have mutually established k as the index of the mobile circle. Writing, the formulae are positioned at will, say such k lies under the upper-case B and the next letter corresponds to the letter that comes next. In writing to you, I will first of all put the upper-case B under which lies the index k in the formula; this is a signal to you far away, wanting to read what I have written, that you should set up the twin formula in your keeping, positioning the mobile circle so that the B sits over the index k. Then all of the rest of the lower-case letters present in the coded text will take their meaning and sound from those of the fixed circle above them.

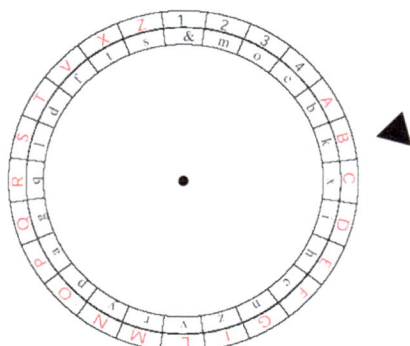

Fig. 5

After I have written three or four words I will mutate the position of the index in our formula, rotating the disk let's say, so that the index k falls below the upper-case R. Then in the missive I write an upper-case R to indicate that k no longer refers to B, but to R, and the letters that follow will assume new meanings.

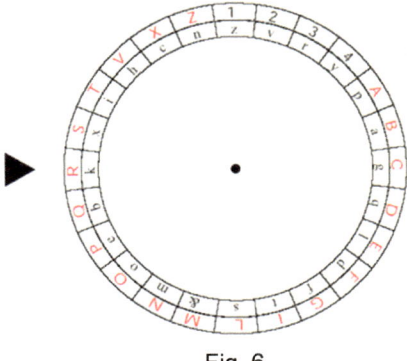

Fig. 6

You likewise, far way and receiving the message, have to look carefully in reading to find the upper-case letter, which you will know serves solely to indicate the positioning of the mobile circle and that the index has changed. Thus you too will position the index under that upper-case letter, and be able to read and understand the entire text with ease.

Chapter XIV

viii. A further image can clarify Alberti's explanation of an alternate use of his *formula*:

Alternately, an index could be selected from among the upper-case letters and we could mutually agree on which would be the index; say we have determined the letter B as the index. The first letter that will appear in the missive that I write to you will be whatever lower-case letter you want, say q; the formula will be positioned so that this lies under the index B.

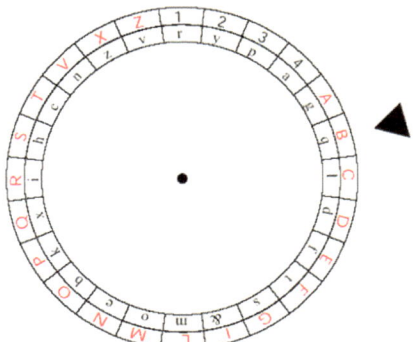

Fig. 7

The description in this chapter confirms that Alberti's coding method is not *monoalphabetic*, but *polyalphabetic*, and this is undoubtedly his original contribution to Western cryptology, a contribution that lay largely unrecognized until the nineteenth century, and which remained undecipherable until that time.

Chapter XVI

ix. The four numerals are used individually to mark a change in the encoded alphabet. The $4 \times 4 = 16$ pairs of numerals, $4 \times 4 \times 4 = 64$ triplets and $4 \times 4 \times 4 \times 4 = 256$ quadruplets are used to set up a table of $16 + 64 + 256 = 336$ encoded phrases. He attributes phrases to each number in tabular form. In reverse, he lists the phrases alphabetically and then shows the numbers.

Chapter XVII

x. The diagram shown in fig. 8 for the *formula* is drawn accurately to Alberti's specification in which the diameter of the *stabilis* is one 1 1/9 of the diameter of the *mobilis*, a ratio of 10:9. The random arrangement of the lowercase characters varies in different manuscripts and editions, but that shown seems to be the most common.

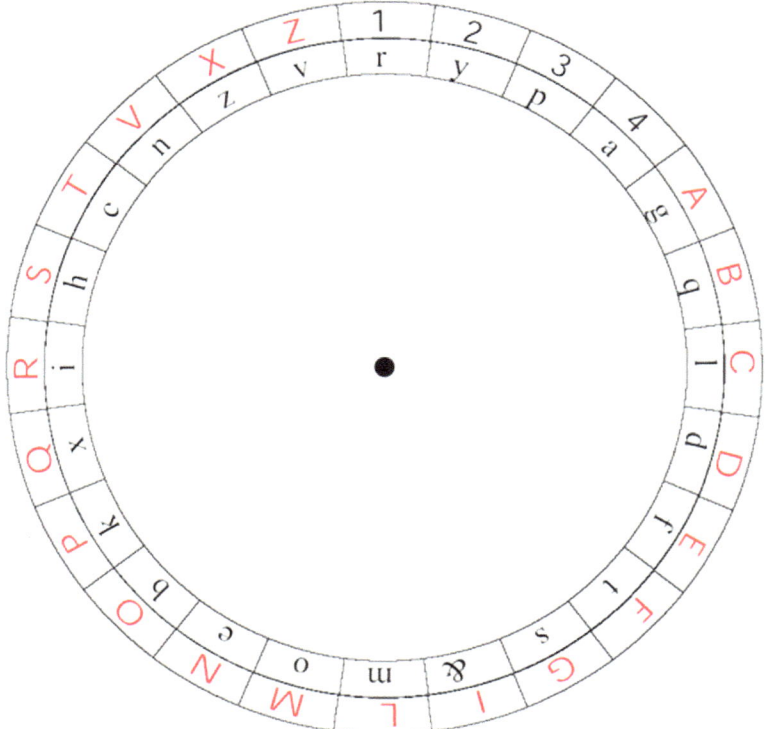

Fig. 8. Alberti's *formula*

Chapter XVIII

xi. To save space, the numerical table on the following page is not according to Alberti's specification, which is a simple list from 11 to 4444 with accompanying plaintext phrases for each of the 336 lines.

11	12	13	14	21	22	23	24
31	32	33	34	41	42	43	44
111	112	113	114	121	122	123	124
131	132	133	134	141	142	143	144
211	212	213	214	221	222	223	224
231	232	233	234	241	242	243	244
311	312	313	314	321	322	323	324
331	332	333	334	341	342	343	344
411	412	413	414	421	422	423	424
431	432	433	434	441	442	443	444
1111	1112	1113	1114	1121	1122	1123	1124
1131	1132	1133	1134	1141	1142	1143	1144
1211	1212	1213	1214	1221	1222	1223	1224
1231	1232	1233	1234	1241	1242	1243	1244
1311	1312	1313	1314	1321	1322	1323	1324
1331	1332	1333	1334	1341	1342	1343	1344
1411	1412	1413	1414	1421	1422	1423	1424
1431	1432	1433	1434	1441	1442	1443	1444
2111	2112	2113	2114	2121	2122	2123	2124
2131	2132	2133	2134	2141	2142	2143	2144
2211	2212	2213	2214	2221	2222	2223	2224
2231	2232	2233	2234	2241	2242	2243	2244
2311	2312	2313	2314	2321	2312	2313	2314
2331	2332	2333	2334	2341	2342	2343	2344
2411	2412	2413	2414	2421	2422	2423	2424
2431	2432	2433	2434	2441	2442	2443	2444
3111	3112	3113	3114	3121	3122	3123	3124
3131	3132	3133	3134	3141	3142	3143	3144
3211	3212	3213	3214	3221	3222	3223	3224
3231	3232	3233	3234	3241	3242	3243	3244
3311	3312	3313	3314	3321	3322	3323	3324
3331	3332	3333	3334	3341	3342	3343	3344
3411	3412	3413	3414	3421	3422	3423	3424
3431	3432	3433	3434	3441	3442	3443	3444
4111	4112	4113	4114	4121	4122	4123	4124
4131	4132	4133	4134	4141	4142	4143	4144
4211	4212	4213	4214	4221	4222	4223	4224
4231	4232	4233	4234	4241	4242	4243	4244
4311	4312	4313	4314	4321	4322	4323	4324
4331	4332	4333	4334	4341	4342	4343	4344
4411	4412	4413	4414	4421	4422	4423	4424
4431	4432	4433	4434	4441	4442	4443	4444

Chapter XIX

xii. A fine summary of the cryptological significance and influence of *De Cifris* is given by David Kahn. An excellent historical and philological review in English is provided by Augusto Buonafalce.

Notes on the Translation of
On Writing in Ciphers

Kim Williams

This present translation of Alberti's treatise on ciphers is based on the Latin text found in the 1998 critical edition of *De componendis cyfris* in Latin compiled by A. Buonafalce with comparisons to made to the 1994 translation by Alessandro Zaccagnini. I determined to undertake the translation because I was struck by the difference in tone between the 1994 Italian translation and the 1997 English translation by the same translator. I am grateful to Monica Ugaglia, Tessa Morrison and Richard Schofield for advice on particularly difficult passages.

ON SQUARING THE LUNE

K. Williams et al. (eds.), *The Mathematical Works of Leon Battista Alberti*,
DOI 10.1007/978-3-0346-0474-1_5, © Springer Basel AG 2010

On Squaring the Lune[1]

The way of measuring a two-cusped[2] figure composed of two curved lines as shown in the figure[3]

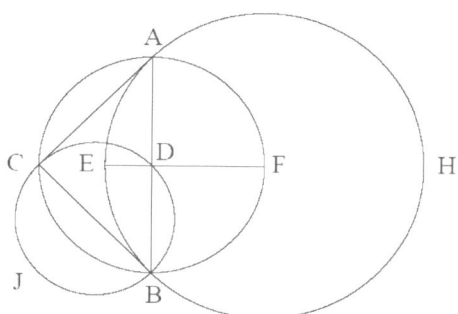

---

[1] This English translation by Kim Williams was based on the version published in G. Mancini's *Leonis Baptistæ Alberti Opera inedita et pauca separatim impressa* [1890: 305-307]; it was checked against the version published by Dominque Raynaud [2006]. In the edition by Mancini, the text was preceded by this Introduction:

Ex codice Florentino bibliothecæ Magliabechianæ 243, classis VI, f.° 77, qui ALBERTI libellum Ludi matematici *inscriptum complectitur. – Hujus problematis solutio desideratur in codicibus Florentinis bibliothecæ Riccardianæ n.° 2110 et n° 2942, nec non in n.° 3 bibliothecæ Morenianæ et in editionibus opuscoli* Ludi matematici *a BARTOLO et BONUCCIO curatis. – Franciscus SIACCI perillustris mathematicus problema revisit et figuræ formam, quæ in codice deerat addere voluit. Problema solutum a Baptista ALBERTO conjicio, sed certissima notitia deest* (Formerly codex 243, class VI, f° 77 in the Magliabecchiana Library in Florence, here included as part of ALBERTI's booklet *Ludi matematici*. The solution to this problem was looked for in Codex no. 2110 and no. 2942 in the Riccardiana library in Florence, and in <codex> no. 3 in the Moreniana library and in the editions of the booklet *Ludi matematici* edited by BARTOLI and BONUCCI. The illustrious mathematician Francesco SIACCI revisited the problem and made the figure, which, missing in the codex, it was desired to add it here. The solution to the problem is attributed to Battista ALBERTO but there is no certain proof of this).

[2] Translator's note: In Italian 'biangolare', bi-angular, or with two cusps.

[3] Translator's note. This figure referred to by Mancini in his Introduction (reproduced on p. 165 in this present volume), was missing in the original codex and was, according to Mancini, created by mathematician Francesco Siacci (Rome 1839 – Naples 1907). A well-known mathematician in his day, Siacci was a professor at the University of Torino, a member of the Accademia delle Scienze and a Senator in the Italian Parliament from 1892. We were unable to locate any publications of Siacci regarding Alberti; it appears that he was asked by Mancini to develop the figure. This present figure and all those that follow are by Lionel March.

Contrary to the opinion of the many who say that figures composed of lines that are curved and circular cannot be squared perfectly, most of all of those that are portions of circles, they say this in my opinion by the authority of Aristotle, who says that *quadratura circuli est scibilis, sed non scita quia est impotentia naturæ*,[4] squaring the circle is knowable though not found but it is in nature's power; and not being able to give the squaring of the circle perfectly, they argue that it is impossible to give the perfect squaring of figures made of curved lines first and foremost circular ones; since I have found the perfect squaring of the figure shown here, that is, a figure with two cusps in the shape of the moon marked AB, I say that if we had had careful investigators, then if squaring the circle is in nature's power, it is likewise in men's power. Thus to demonstrate the squaring of said figure AB, after first noting two propositions of Euclid pertaining to the declaration, I will tell the way it is done.

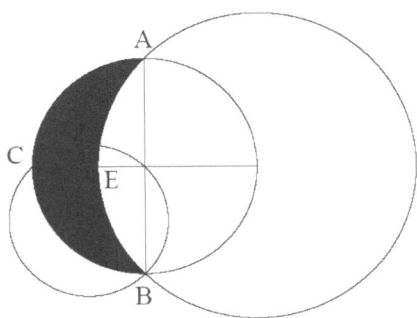

First proposition. In Book XII, proposition 2:
Circles are to one another as the squares on the diameters[5]
Proposition in Book II, no. 47[6]
In right-angled triangles the square on the side subtending the right angle is
equal to the squares on the sides containing the right angle.

---

[4] The biggest discrepancy between the Mancini text and the Raynaud version: Mancini, *impotentia naturae*; Raynaud, *in potentia nacture*.
[5] Translator's note. [Euclid 1956: vol. 3, 371]. The translation of this proposition and the next were taken directly from Heath's translation of Euclid.
[6] Translator's note. [Euclid 1956: vol. 1, 349]. The Italian original from Mancini calls this proposition 46, and the note says, 'In the codex of Euclid that was of [owned by? used by?] Alberti and is now in the Marciana library in Venice (latin. 39, classis VIII), this proposition is no. 46, book I, fol. 9'. We verified this, and it is also confirmed by Raynaud 2006.

I say that the squaring of the lune ABEC[7] will be proper to the surface equal to the triangle ABC inscribed in the half-circle,

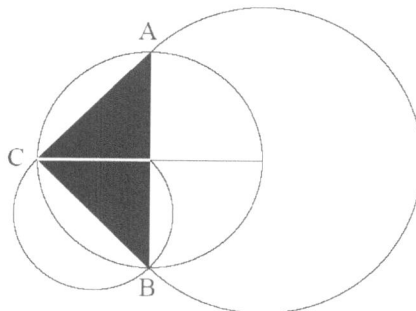

which triangle contains the two parts marked AE[D] and BD[E]

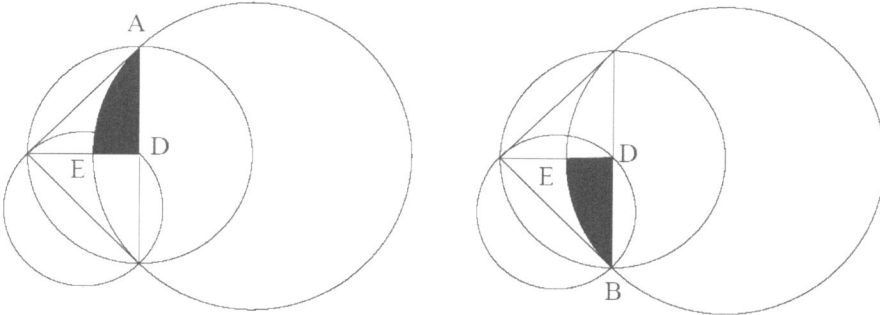

that are portions of the marked[8] circle, which two parts are equal to the two portions of circle AC and BC by the second <proposition>[9] of Euclid's <book> XII cited above and by <proposition> 47 of <book> 1.

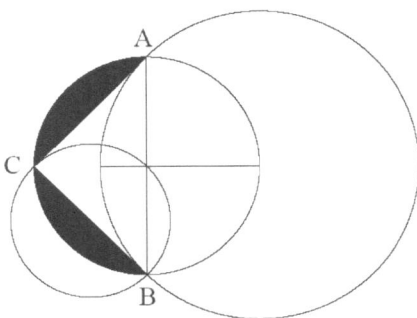

The first proposition cited here manifestly shows that it there is a double proportion between circle ABCF and circle ABEH,

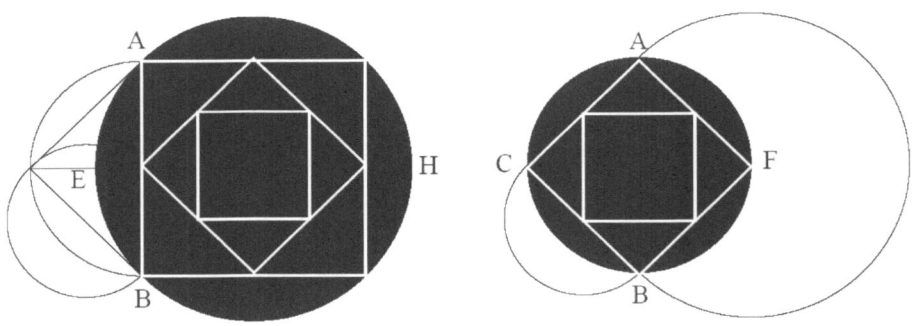

---

[8] Translator's note: In the ms. is an abbreviation that is interpreted as 'sig.le' or 'singolare' (singular) by Mancini and as 'sig.te' or 'marked, signed' by Raynaud. We agree that 'marked' makes more sense in the context of explanation.
[9] Translator's note. This and all further additions to the text made for clarification are marked in angle brackets < >.

because the side of the square containing the larger circle is the diameter of the other, second, square, and here again it happens that [proposition] 47 of <book> I which manifestly shows that they are in double proportion and the side of the square placed in the second circle is the diameter of the smaller circle, that is, BCJD,

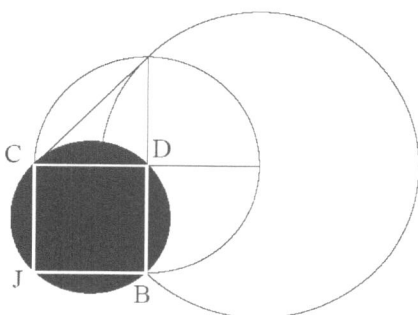

that thus they go on proportioning themselves, and always in double proportion: it thus follows that the squares placed in the circles are also in double proportion as can be seen that they must, and thus in like fashion the portions of circles are in double <proportion>. Therefore two smaller portions make a greater, that is, the portions AC and BC joined together are equal to the portion ABDE,

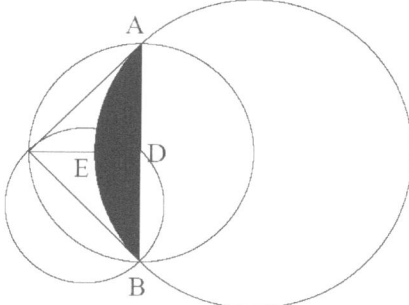

which was the proposition:[10] and in forming triangle ABC comes, in the place of the two above-mentioned portions AC and BC, the portion of the larger triangle, that is, ABED, which is equal to the two smaller ones. Manifestly thus it is seen that triangle ABC is precisely equal to the lune,

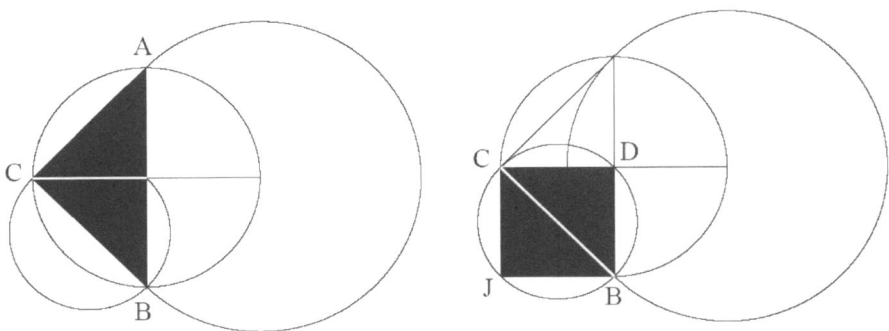

then because of this, from this square figure we can argue that as the square of this lune composed of two curved lines was found, it is likewise possible to square the circle.

Lionel March

This 'Alberto' paper is attributed to Alberti, but there is no definite proof of this.[1] There are contextual reasons for supposing that Alberti might be the author. The classical problem of squaring the circle,[2] or the search for a value of π, was a preoccupation of mathematicians in Italy during the fifteenth century.[3] A leading investigator was Nicholas Cusanus (1401–1464). It is not known explicitly whether Alberti (1404–1472) was an acquaintance of Cusanus, but he would undoubtedly have been familiar with his work. It is known, from a letter addressed by the distinguished mathematician Regiomontanus (Johannes Muller, 1436–1476), that Alberti was a significant astronomical collaborator with Paolo Toscanelli (1397–1482).[4] A polymath, Toscanelli is most famously known among architects for his engineering contributions to Florence's Duomo. Alberti dedicated his *Intercenales* (*Dinner Pieces*), to Toscanelli: 'My dear Paolo, continue to love your dear friend Leon Battista ... mindful of our long friendship'.[5]

Nicholas Cusanus probably first met Toscanelli at Padua University, Paul Lawrence Rose writes:

> Later enthused with admiration for Archimedes, Cusanus wrote several works on squaring the circle, dedicating his *De Geometricis Transmutationibus* and *De Arithmeticus Complementis* to the Florentine. In return, Toscanelli addressed a short mathematical treatise or letter to Cusanus in 1453–1454, and soon afterwards was enlisted by Cusanus as an interlocutor in the cardinal's *Dialogus inter Cardinalem Sancti Petri et Paulum Physicum de Circuli Quadratura* (1457).

Toscanelli is less than truthfully represented in the *Dialogus*. Yet later in 1471, Regiomontanus, who tackled the quadrature of the circle himself, declared Cardinal

---

[1] But see Raynaud [2006] who concludes that the attribution to Alberti is incorrect.

[2] E. W. Hobson [2007, 1913] gives an excellent, brief account of the history of squaring the circle.

[3] Paul Lawrence Rose [1975] sets a humanistic context for the quadrature of the circle within the mathematics of the Italian Renaissance.

[4] J. V. Field [2005] relates the mathematical works of Nicholas Cusanus and Regiomontanus to Piero della Francesca (ca. 1415–1492) and his remarkable geometric and algebraic investigations. In passing, she notes that a manuscript of Alberti's *Elementi di pittura* was in Cusanus's library.

[5] Dedication to Toscanelli comes from David Marsh [1987: 15].

Cusanus to be *geometra ridiculus* while continuing to hold Toscanelli in the highest regard.

There were essentially two approaches to quadrature. The first recognized that the areas of inscribed and circumscribed polygons converge on the area of the circle as the number of sides increased, most commonly 6, 12, 24, 48, 96 sides. The second was followed by Alberti. Show that one curvilinear shape can be squared, and trust that the quadrature of the circle follows. The ancient Greek lune of Hippocrates[6] (ca. 460 BCE) is the example chosen, although the author of the 'Alberto' paper does not acknowledge this source. Nevertheless, despite the mathematical competence of the proof, the author makes the terrible logical error in which a special case, the quadrature of a particular lune, implies the successful outcome of a different case, the possibility of the quadrature of the circle. Alberti had the skills for the proof, but is it conceivable that he would make such an elementary logical error? Or has he copied an extant argument out of interest for his own record, leaving the fallacious conclusion unchallenged?[7]

It has been known since the mid-1700s that the lune chosen is one of only five that can be squared by Euclidean straight edge and compass methods. Heath gives a parametric, trigonometric equation: $\sin mA = \sqrt{m}.\sin A$ where angle 2A suspends the inner arc of the lune and the equation reduces to a quadratic (to satisfy the Euclidean condition). Five values of m satisfy the conditions: m = 2, 3, 3/2, from top to bottom left, and 5, 5/3 from top to bottom right. It is generally agreed that Hippocrates knew of three such lunes (left hand column).

---

[6] Sir Thomas Heath [1981, 1921] gives an authoritative account of the lunes of Hippocrates of Chios.
[7] Joan Gadol suggests that *De lunularum quadratura* may be related to calibration problems in the apparatus described in Alberti's *De Statua* [1969: 78].

The squaring of the circle absorbed some of the best minds of the Italian Renaissance. At the turn of the sixteenth century, Leonardo da Vinci (1452–1519) 'was so obsessed with mathematics that he neglected his painting and, writes an observer, "the sight of a brush puts him out of temper"'.[8] On 30 November 1504, Leonardo reported in the margin of folio 112 recto in Codex Madrid II: 'The night of St. Andrew, I finally found the quadrature of the circle …'.[9] But he had not, and as was shown later by Ferdinand von Lindemann (1852-1939) in 1882, no Euclidean construction is possible to give the transcendental value π.[10]

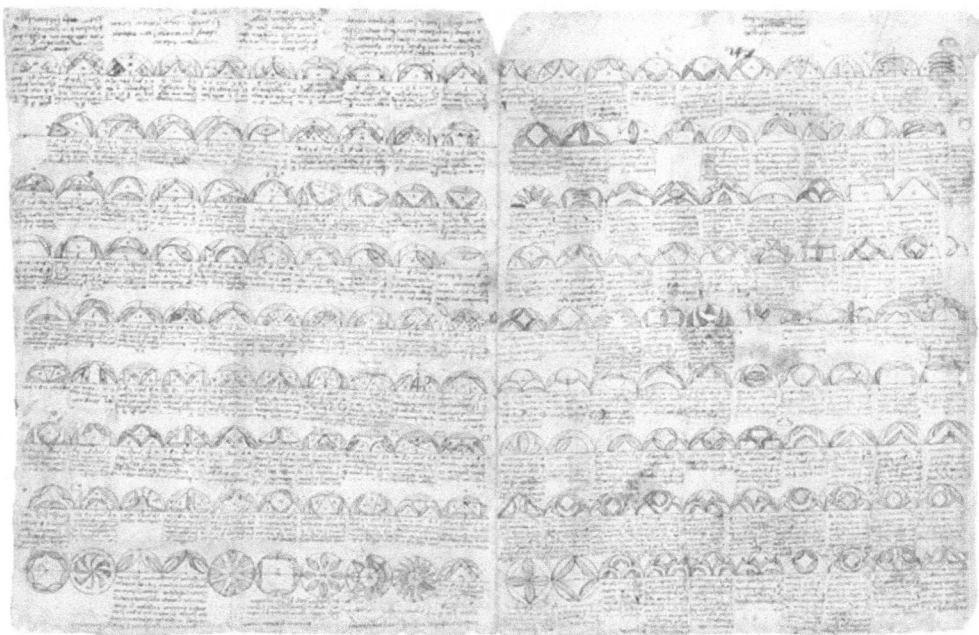

Leonardo da Vinci, studies of lunes, *Codex Atlanticus* fol. 455r. 'The abstract game of squaring the circle comes to life under Leonardo's pen in rows of crescents, paired leaves, pinwheels, and rosettes' [Marinoni 1990: 77]

The quadrature of the circle was a quest for a value of π. Alberti uses 22/7 as rational approximation for π. He does so explicitly in *Ludi*,[11] and implicitly in *De re aedificatoria*. In Book 7 [1988: 219], Alberti discusses the interior height to diameter dimensions of circular temples. After exploring the ratios half (1:2), two thirds (2:3), and three quarters (3:4), the ratio of 'eleven to four' is recommended for 'the more experienced'. Either a careful calculation following Alberti's text, or a simple drawing shows how wrong this

---

[8] [Marinoni 1990: 70].
[9] Augusto Marinoni [1990]) mentions Leonardo's search for the quadrature of the circle and proposes that the drawings from Codex Atlanticus are closely related.
[10] [Hobson 2007: 51–53]
[11] See pp. 36–37 and 102 in this present volume.

proportion is. Indeed, the calculation shows that, using 22/7 for π, the correct ratio is 11:14. This ratio may be interpreted as either the circumference of the inscribed circle to the perimeter of the square, or the area of the inscribed circle to the area of the square: a different take on 'squaring the circle'

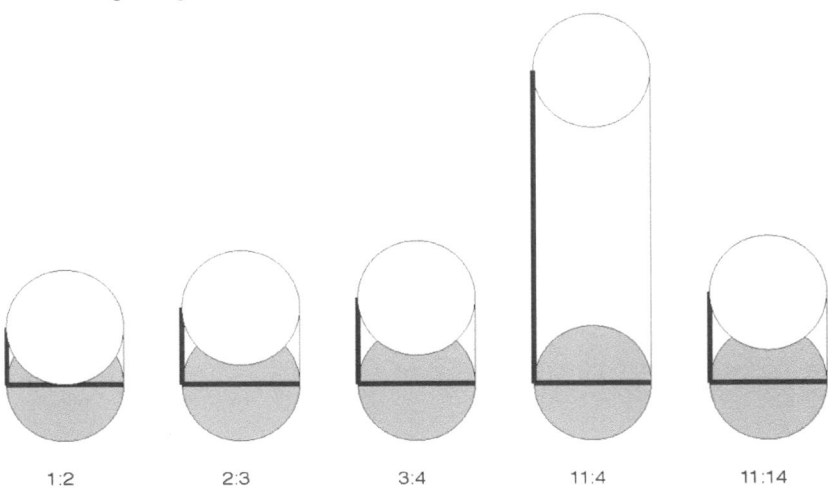

| 1:2 | 2:3 | 3:4 | 11:4 | 11:14 |

Just preceding this, Alberti [1988: 218 (VII, ix)] looks at appropriate arrangements for Doric intercolumniations. The column is in proportion seven to one, height to diameter. A four column Doric elevation is divided into 'twenty-seven parts': 2 + 6 + 2 + 7 + 2 + 6 + 2, where the column diameter is 2 and the middle space is one sixth larger than the side intercolumniations.

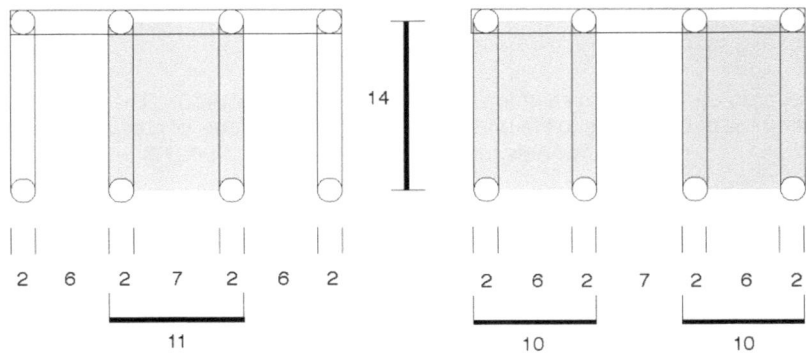

Again, a 14 : 11 ratio is found, accompanied by two 7 : 5 (square root 2) ratios on either side. In such a context, this suggests that 14 : 11 might also be interpreted as a rational convergent to the cube root of 2, in the spirit of Alberti's *De re aedificatoria*, IX, vi [Alberti 1988: 307; see also March, 1996, 1999].

Bibliography

Primary Sources

Alberti, Leon Battista. *Ex ludis rerum mathematicarum*

Manuscripts (existence verified at the time of this printing):
    Cambridge, MA: Houghton Library Ms. Typography 316
    Cambridge, MA: Houghton Library Ms. Typography 422/2
    Florence: Biblioteca Riccardiana 2110
    Florence: Biblioteca Riccardiana 2942, 1
    Florence: Biblioteca Nazionale Centrale Galileiana 10, fols. 1r-16r
    Florence: Magliabechiana IV 243
    Florence: Biblioteca Laurenziana Ashburnham 356
    Florence: Biblioteca Moreniana 3
    Genova: Biblioteca Universitaria di Genova G. IV. 29
    Ravenna: Biblioteca Classense 208
    Rome: Biblioteca Nazionale Centrale Fondo "Vittorio Emmanuele III", no. 574
    Rouen: Bibliothèque Municipale Leber 1158 (3056)
    Venice: Biblioteca Nazionale Marciana, it. XI. 067 (7351)

Publishing history:

1568. *Delle piacevolezze delle matematice.* In *Opuscoli morali di L. B. Alberti gentil'huomo fiorentino ... Tradotti et parte corretti da M. Cosimo Bartoli.* Venice: Franceschi, pp. 225–255.

1847. *Ludi matematici.* In *Opere vulgari di Leon Battista Alberti*, Anicio Bonucci, ed. vol. IV, pp. 405–440. Florence: Tipografia Galileiana.

1926. *Ludi matematici.* Sebastiano Timpanaro. Pp. 1-36 in *Leonardo.* No. 1 in the book series *Pagine di scienza* Milan: Mondadori (Edizioni Mondadori per le scuole medie).

1973. *Ludi rerum mathematicarum.* In *Leon Battista Alberti, Opere Volgari.* C. Grayson, ed. Bari: Laterza, vol. III, pp. 131–173

1980. *Ludi matematici.* Raffaele Rinaldi, ed. Preface by Ludovico Geymonat. Milan: Guanda Editore.

Contemporary translations

1937. *Matematicheske zabavi* (Russian translation). V. P. Zubow, trans. In vol. II of *Desyat knig o zodchestve* (*De re aedificatoria*), pp. 67-100. Moscow.

2002. *Divertissements mathématiques* (French translation). Pierre Souffrin, trans. Paris: Seuil.

Alberti, Leon Battista. *Elementa picturae / Elementi di pittura* (*Elements of Painting*)

Manuscripts (existence verified at the time of this printing):
Cambridge, MA: Houghton Library Typ. 422/1
Vatican City: Biblioteca Apostolica Vaticana, Vat. Lat. 3151 (Latin)
Lucca: Biblioteca Governativa, Ms. 1448 (Latin)
Vatican City: Biblioteca Apostolica Vaticana, Vat. Lat. 8104 (Latin)
Vatican City: Biblioteca Apostolica Vaticana, Ott. Lat. 1424 (Latin)
Florence: Biblioteca Nazionale Centrale, ms. II.IV.039 (was Magl. XVII,6)
(Latin)
Florence: Biblioteca Riccardiana 0927 (Latin)
Milan: Biblioteca Ambrosiana, ms. O. 80. Sup.
Oxford: Bodleian Library, Canon. Misc. 121 (Latin)
Oxford: Bodleian Library, Canon. Misc. 172
Paris: Bibliothèque Nationale, lat. 10252 (Latin)
Verona: Biblioteca Capitolare, CCLXXIII (vulgar)
Bernkastel-Kues, St. Nikolaus-Hospitals Cusanusstift, cod. 112.

Publishing history
1864. *Gli elementi di pittura per la prima volta pubblicati con un discorso sulla parte avuta dall'Alberti nel rimettere in onore la lingua italiana nel secolo XV*. Girolamo Mancini, ed. Cortona: Bimbi.

1890. *Elementa picturae/Elementi di pittura*. In: *Opera inedita et pauca separatim impressa*, G. Mancini, ed. Florence: Sansoni, pp. 47–65 (Latin and Italian texts on facing pages).

1973. *Elementi di pittura*. In *Leon Battista Alberti, Opere Volgari*. C. Grayson, ed. Bari: Laterza, vol. III, pp. 109–129 (Latin and Italian texts on facing pages).

Contemporary translations:
2000. *Das Standbild, die Malkunst, Grundlagen der Malerei – De pictura, De statua et Elementa Picturae*, Christoph Schaüblin, trans. Introduction and commentary by Oskar Bätschmann. Darmstadt: Wissenschaftliche Buchgesellschaft.

Alberti, Leon Battista. *De componendis cifris* (*On Writing in Ciphers*)

Manuscripts (existence verified at the time of this printing):
Florence: Firenze, Biblioteca Nazionale Centrale, Cod. II, IV, 39, Fondo Principale (Magliabechianus II IV 39)
Florence: Biblioteca Biblioteca Riccardiana 0767
Florence: Biblioteca Marucelliana ms. B. VI. 39
Florence: Biblioteca Biblioteca Riccardiana, 0927
Paris: Bibliothèque Nationale, lat. 8754
Vatican City: Biblioteca Apostolica Vaticana MS Chigi M II 49, vol. 35
Vatican City: Vatican Library Cod. Vaticanus Latinus 6532
Vatican City: Biblioteca Apostolica Vaticana Cod. Vaticanus Latinus 5118

Vatican City: Varia Politicorum LXXX
Vatican City: Vatican Library Cod. Vaticanus Latinus 5357
Venice: Archivio di Stato Venezia CCX VI 1
Venice: Biblioteca Nazionale Marciana Cod. Marc. Lat. XIV 32
Venice: Museo Civico Correr, Miscellanea Correr 47 2130

Publishing history

1568. *La cifra*. In *Opuscoli morali di L. B. Alberti gentil'huomo fiorentino ... Tradotti et parte corretti da M. Cosimo Bartoli.* Venice: Franceschi, pp. 200–219.

1890. *De cifra.* In: *Opera inedita et pauca separatim impressa*, G. Mancini, ed. Florence: Sansoni, pp. 310 (Proemio only).

1906. *Die Geheimschrift im Dienste der päpstlichen Kurie von ihren Anfängen bis zum Ende des XVI. Jahrhunderts.* Aloys Meister, ed. Schöningh: Paderborn, pp. 125–141.

Contemporary translations

1994. *Dello scrivere in cifra.* (Italian translation). M. Zanni, trans. Torino: Galimberti Tipografi.

1997. *A treatise on ciphers* (English translation). Alessandro Zaccagnini, trans. Torino: Galimberti Tipografi.

1998. *De componendis cyfris* (Critical edition). Augusto Buonafalce, ed. Galimberti Tipografi Editori, Torino.

2000. *De cifris* (French translation). Martine Furno, trans. In *Leon Battista Alberti: Actes du Congrès International de Paris* (Sorbonne-Institut de France-Institut culturel italien-Collège de France, 10–15 April 1995). Paris: Librairie Philosophique J. Vrin & Torino: Nino Aragno Editore, vol. II, pp. 705–725.

Alberti, Leon Battista. *De lunularum quadratura* (*On Squaring the Lune*)
Manuscripts (existence verified at the time of this printing):
Florence: Biblioteca Nazionale Centrale, Magliabechiana IV 243

Publishing history

1890. *De lunularum quadratura.* In: *Opera inedita et pauca separatim impressa*, Girolamo Mancini, ed. Florence: Sansoni, pp. 305–307.

1978. The *De lunularum quadratura* attributed to Leon Battista Alberti. Pp. 1326–1328 In *The Fate of the Medieval Archimedes 1300-1565*, vol. III of Archimedes in the Middle Ages, Marshall Clagett, ed. Philadelphia: American Philosophical Society.

2006. Le traité sur la quadrature des lunules attribué à Leon Battista Alberti. Dominique Raynaud. *Albertiana* 9: 65 & 67.

Contemporary translations

2006. Le traité sur la quadrature des lunules attribué à Leon Battista Alberti. Dominique Raynaud. *Albertiana* 9: 64 & 66.

Secondary sources

ALBERTI, Leon Battista. 1844-1850. *Opere vulgari di Leon Battista Alberti*, Anicio Bonucci, ed. 5 vols. Florence: Tipografia Galileiana.

ALBERTI, Leon Battista. 1890. *Opera inedita et pauca separatim impressa*. G. Mancini, ed. Florence: Sansoni.

ALBERTI, Leon Battista. 1960–1973. *Opere volgari*. Cecil Grayson, ed. 3 vols. Vol. 1, 1960 (Scrittori d'Italia 218); Vol. 2, 1966 (Scrittori d'Italian 234); Vol. 3, 1973 (Scrittori d'Italia 254). Bari: Laterza.

ALBERTI, Leon Battista. 1987. *Dinner Pieces*. David Marsh, trans. Birmington, New York: The Renaissance Society of America.

ALBERTI, Leon Battista. 1988. *On the Art of Building in Ten Books*. Joseph Rykwert, Neil Leach and Robert Tavernor, trans. Cambridge MA: MIT Press.

ALBERTI, Leon Battista Alberti. 2004a. *On Painting*. Cecil Grayson, trans. Introduction and notes by Martin Kemp. Rpt. with revised Further Reading, London: Penguin Classics. (1st edition, London: Phaidon Press 1991).

ALBERTI, Leon Battista. 2004b. *La Peinture. Texte Latin, Traduction Francaise, Version Italienne*. Thomas Golsenne and Bertrand Prévost, eds. Introduction by Yves Hersant. Paris: Éditions du Seuil.

ALBERTI, Leon Battista. 2007. *Delineation of the City of Rome (Descriptio urbis Romae)*. Mario Carpo and Francesco Furlan, eds. Tempe AZ: Arizona Center for Medieval and Renaissance Studies.

ANDERSEN, Kirsti. 2007. *The Geometry of an Art: The History of the Mathematical Theory of Perspective from Albert to Monge*. New York: Springer.

ARRIGHI, Gino. 1972. I "Ludi matematici" di Leon Battista Alberti. Pp. 31-49 in *Civiltà dell'Umanesimo: Atti del VI°, VII°, VIII° Convegno internazinale del Centro di Studi Umanistici* (Montepulciano, Palazzo Tarugi, 1969, 1970, 1971). Giovannangiola Tarugi, ed. Florence: Olschki.

BELLI, Silvio. 1565. *Del misurare con la vista*. Venice.

BELLI, Silvio. 1573. *Della Proportione, et Proportionalità, Communi Passioni del Quanto, Libri Tre*. Venice.

BELLI, Silvio, 2003. *On Ratio and Proportion, The Common Properties of Quantity*. English trans. and commentary Stephen R. Wassell and Kim Williams, with a foreword by Lionel March. 2nd ed. Fucecchio, Florence: Kim Williams Books.

CALINGER, Ronald. 1996. *Vita Mathematica: Historical Research and Integration with Teaching*. Washington, D.C.: Mathematical Association of America.

D'AMORE, Bruno. 2005. Leon Battista Alberti ed i suoi *Ludi rerum mathematicarum*. *Il Carobbio* **XXX**: 61–66.

DODDS, George and Robert TAVERNOR, eds. 2002. *Body and Building: On the Changing Relation of Body and Architecture*. Cambridge, MA, and London: MIT Press.

EUCLID. 1956. *The Thirteen Books of Euclid's Elements*. 3 vols. (2nd unabridged ed.). Thomas L. Heath, trans. rpt. New York: Dover Publications.

EUGENI, Franco and Fabio MERCANTI. 2004. Leon Battista Alberti e Blaise de Vigenère: dai *Ludi Matematici* agli arbori della moderna crittografia. Quarto Congresso Società Italiana per la Storia della Matematica (SISM), Padova, 9–11 September 2004, Sunti (unpublished).

FIBONACCI. 2008. *Fibonacci's De Practica Geometrie*. Sources and Studies in the History of Mathematics and Physical Sciences **XXXVI**. Hughes, Barnabas, ed. (1st ed. 2007).

FIELD, J. V. 2005. *Piero della Francesca: A Mathematician's Art*. New Haven and London: Yale University Press.

FILARETE (Antonio Averlino). 1972. *Trattato di Architettura*. Anna Maria Finoli and Liliana Grassi, eds. Milan: Edizioni il Polifilo.

FURLAN, Francesco. 2001. 'Ex Ludis Rerum mathematicarum': appunti per un'auspicabile riedizione critica. Pp. 147–165 in *Mélange en l'honneur de Charles Bec*, F. Livi, ed. Paris: Presses de la Sorbonne.

FURLAN, Francesco. 2006. In margine all'edizione degli *Ex ludis rerum mathematicarum:* Ossia osservazioni e note per l'edizione di un testo scientifico e delle sue figure. *Revue d'histoire des sciences* 59-2 (July-December 2006): 197–217.

FURLAN, Francesco and Pierre SOUFFRIN. 2001. Philologie et histoire des sciences: le Problème XVIIe des 'Ludi Rerum Mathematicarum'. *Albertiana* 4: 3–32.

GADOL, Joan. 1969. *Leon Battista Alberti: Universal Man of the Early Renaissance*. Chicago: University of Chicago Press.

GAMBUTI, Alessandro. 1972. Nuove ricerche sugli Elementa Picturae. *Studi e documenti di architettura* 1 ('Omaggio ad Alberti'): 131–171.

GRAFTON, Anthony. 2000. *Leon Battista Alberti: Master Builder of the Italian Renaissance*. New York: Hill and Wang.

GRAYSON, Cecil. 1954. La Prima Edizione del "Philodoxeos". *Rinascimento* 5, 1: 291-293.

GRAYSON, Cecil. 1960. The composition of L.B. Alberti's *Decem Libri de re aedificatoria*. *Müncher Jahrbuch der bildenen Kunst* III, ii: 161 ff.

GRAYSON, Cecil. 1998. *Studi su Leon Battista Alberti*. Paolo Claut, ed. Florence: Leo S. Olschki.

GREENSTEIN, Jack M. 1992. *Mantegna and painting as historical narrative*. Chicago and London: The University of Chicago Press.

GUIDONE, Mario. 2001. Ostilio Ricci da Fermo. Un ponte tra Galileo e la scienza rinascimentale. In *Scienziati e tecnologi marchigiani nel tempo. Quaderni del Consiglio Regionale Delle Marche* V, 30 (February 2001): 59-74

HEATH, Thomas. 1921. *A History of Greek Mathematics*. Oxford: Clarendon Press. Rpt. Dover Edition, 1981.

HOBSON, E. W. 1913. *Squaring the Circle: a History of the Problem*. Cambridge: Cambridge University Press. Rpt. Milton Keynes: Merchant Books, 2007.

HUGHES, Barnabas, ed. 2008. *Fibonacci's De practica geometrie*. Sources and Studies in the History of Mathematics and Physical Sciences. New York: Springer.

KAHN, David. 1967. *The Codebreakers: The Comprehensive History of Secret Communication from Ancient Times to the Internet*. New York: Scribner (revised and updated 1996).

KATZ, Victor J. 1998. *A History of Mathematics: An Introduction*. 2nd ed. New York: Addison Wesley Longman.

LANDAU, David and Peter W. PARSHALL. 1994. *The Renaissance Print: 1470–1550*. New Haven CT: Yale University Press.

LÉVY-LEBLOND, Jean-Marc. 2003. Columella's Formula. *The Mathematical Intelligencer* 25: 51–54.

MANCINI, Girolamo. 1882. *Vita di Leon Battista Alberti*. Florence: Sansoni.

MANCINI, Girolamo. 1917. Leon Battista Alberti. Pp. 21-41 in *Vite cinque annotate da Girolamo Mancini*. Florence: Stab. Tipografia G. Carnesecchi e Figli.

MARASCHIO, Nicoletta. 1972. Aspetti del bilinguismo albertiano nel *De pictura*. Rinascimento 2d series 12: 183–228.

MARCH, Lionel. 1998. *Architectonics of Humanism: Essays on Number in Architecture*. West Sussex: Academy Editions.

MARCH, Lionel. 1996. Renaissance mathematics and architectural proportion in Alberti's *De re aedificatoria*. *Architectural Research Quarterly* 2, 1: 54-65.

MARCH, Lionel. 1999. Proportional design in L. B. Alberti's *Tempio Malatestiano*, Rimini. *Architectural Research Quarterly* 3, 3: 259-269.

MARINONI, Augusto. 1990. The writer. Leonardo's literary legacy. In *The Unkown Leonardo*. Ladislao Reti, ed., New York: Abradale.

MASSALIN, Paola and Branko MITROVIĆ. 2008. Alberti and Euclid / L'Alberti ed Euclide. *Albertiana* 9: 165–249.

MERCANTI, Fabio and Paola LANDRA. 2007. I 'Ludi matematici' di Leon Battista Alberti. *EIRIS* 2: 15–47.

MIDDLETON, W. E. Knowles. 1969. *Invention of the Meteorological Instruments*. Baltimore, Maryland: The John Hopkins Press,

MITROVIĆ, Branko. 2001. A Palladian Palinode: Reassessing Rudolf Wittkower's *Architectural Principles in the Age of Humanism*. architectura **31**: 113–131.

MITROVIĆ, Branko. 2010. Studying Renaissance Architectural Theory in the Age of Stalinism.

MITROVIĆ, Branko and Stephen R. WASSELL. 2006. *Andrea Palladio: Villa Cornaro in Piombino Dese*. New York: Acanthus Press.

MONTI, D. V. 2005. *Breviloquium: Works of St. Bonaventure*, Vol IX. Saint Bonaventure University.

NICCOLINI, Giovanni Batista. 1819. *Elogio di Leon Batista Alberti composto da Gio. Batista Niccolini Segretario dell'Imp. e R. Accademia delle Belle Arti di Firenze e letto da esso nel giorno della solenne distribuzione dei premi maggiori l'anno 1819*. Florence: Niccolò Carli.

PALLADIO, Andrea. 1997. *The Four Books of Architecture*. Robert Tavernor and Richard Schofield, trans. Cambridge MA: MIT Press.

PASQUALE, Salvatore Di. 1992. Tracce di statica archimedea in Leon Battista Alberti. *Palladio* 9 (Gennaio–Giugno 1992): 41–68.

PRAGER, Frank D. and Gustina SCAGLIA. 1972. *Mariano Taccola and His Book* De Ingeneis. Cambridge MA: MIT Press.

PROCISSI, Angiolo. 1959. *La collezione Galileiana nella Biblioteca nazionale di Firenze*, vol. I, Rome: Istituto Poligrafico dello Stato.

RAYNAUD, Dominique. 2006. Le traité sur la quadrature des lunules attribué à Leon Battista Alberti.. *Albertiana* 9: 31–67.

RICHTER, Jean Paul, ed. 1970. *The Notebooks of Leonardo da Vinci*. Mrs. R. C. Bell, trans. 2 vols. New York: Dover Publications.

ROSE, Paul Lawrence. 1975. *The Italian Renaissance of Mathematics: Studies of Humanists and Mathematicians from Petrarch to Galileo*. Geneva: Libraire Droz.

RYKWERT, Joseph and Anne ENGEL, eds. 1994. *Leon Battista Alberti*. Milan: Olivetti/Electa.

SETTLE, Thomas B. 1971. Ostilio Ricci, a Bridge between Alberti and Galileo. Pp. 121-125 in vol. III/B, *Actes du XIIᵉ Congrès International d'Histoire des Sciences (Paris 1968)*. Paris: Blanchard.

SIMONELLI, Maria Picchio. 1971. On Alberti's Treatises of Art and their Chronological Relationship. *Yearbook of Italian Studies* **1**: 75–102.

SOUFFRIN, Pierre. 1998. La geometria practica dans les *Ludi rerum mathematicarum*. *Albertiana* 1: 87–104.

TAVERNOR, Robert. 1998. *On Alberti and the Art of Building*. New Haven-London: Yale University Press.

TAVERNOR, Robert. 2007. *Smoot's Ear: The Measure of Humanity*, New Haven and London: Yale University Press.

VAGNETTI, Luigi. 1972. Considerazioni sui Ludi Matematici. *Studi e documenti di architettura* **1** ('Omaggio ad Alberti'): 173–259.

VASARI, Giorgio. 1991. Vita di Leon Batista Alberti Architetto Fiorentino. Pp. 389-392 in *Le vite dei più eccellenti pittori, scultori e architetti*. Rome: Newton Compton editore.

VILLARD DE HONNECOURT. 2009. *The Portfolio of Villard de Honnecourt: (Paris, Bibliothèque nationale de France, MS FR 19093) A New Critical Edition and Color Facsimile*. Carl F. Barnes Jr., ed. Surrey: Ashgate.

VITRUVIUS. 1584. *I Dieci libri dell'architettura di M. Vitruvio, tradotti e commentati dal Monsig. Daniel Barbaro eletto Patriarca di Aquileia, da lui riveduti e ampliati, & hora in forma più commoda ridotta.* Venice: Francesco de' Franceschi Senesi.

VITRUVIUS. 1960. *The Ten Books on Architecture.* Morris Hickey Morgan, trans. New York: Dover Publications.

VITRUVIUS. 2009. *On Architecture.* Richard Schofield, trans. Introduction by Robert Tavernor. London: Penguin Classics.

WASSELL, Stephen R. 2002. Rediscovering a Family of Means *The Mathematical Intelligencer* **24**, 2: 58–65.

WITTKOWER, Rudolf. 1998. *Architectural Principles in the Age of Humanism.* Chichester, West Sussex: Academy Editions.

ZUPKO, Ronald Edward. 1981. *Italian Weights and Measures from the Middle Ages to the Nineteenth Century.* Philadelphia: American Philosophical Society.

About the authors

Kim Williams is an independent scholar, translator and publisher. She is the editor-in-chief of the *Nexus Network Journal* for architecture and mathematics (Basel: Birkhäuser, 1999-present), and director of the biennial Nexus conferences for architecture and mathematics (1996-present). She is author of *Italian Pavements: Patterns in Space* (Anchorage Press, 1997), *The Villas of Palladio* (Princeton Architectural Press, 2003), and collaborator with Lionel March and Stephen Wassell on the translation/commentary of *On Ratio and Proportion* by Silvio Belli (Kim Williams Books, 2003). She recently co-edited *Discoveries in the History of Mechanics 1600-1800. Essays by David Speiser* (Birkhäuser, 2008). As a translator, she specializes in mathematics, architecture, engineering, and history of the sciences. She is the translator of *Mathematical Lives*, edited by C. Bartocci, R. Betti, A. Guerraggio, and R. Lucchetti (in preparation, Springer-verlag, 2010).

Lionel March holds a BA (Hons) in Mathematics and Architecture, Diploma in Architecture, MA, and Doctor of Science (ScD) from the University of Cambridge. He has held fellowships with the Royal Society of Arts, the Institute of Mathematics and its Applications, and the Royal College of Art. He was founding director of the Centre for Land Use and Built Form Studies, now the Martin Centre for Architectural and Urban Studies, University of Cambridge. He has held Professorships in Engineering (Waterloo, Ontario), Technology (The Open University), Architecture and Urban Design, and Design and Computation (University of California, LA). He was Rector and Vice-Provost (Royal College of Art, London). He was co-editor with Sir Leslie Martin of the twelve volume Cambridge Architectural and Urban Studies, Cambridge University Press. He was founding editor in 1967 of the bi-monthly refereed journal Environment and Planning B, now Planning and Design.
Relevant recent publications include *Architectonics of Humanism: Essay on Number in Architecture* (1999); 'Renaissance mathematics and architectural proportion in Alberti's *De Re Aedificatoria*', *Architectural Research Quarterly* (*ARQ*), 2/1, 1996, 54-65; 'Proportional design in L. B. Alberti's Tempio Malatestiano, Rimini', *ARQ*, 3/3, 1999, 259-269; 'Palladio, Pythagoreanism and Renaissance Mathematics', *Nexus Network Journal*, 10/2, 2008, 227-243. Lionel March is currently Visiting Scholar, Martin Centre for Architectural and Urban Studies, Department of Architecture, University of Cambridge, and emeritus member, Center for Medieval and Renaissance Studies, University of California, Los Angeles.

Stephen R. Wassell holds a B.S. in Architecture, a M.S. and Ph.D. in Mathematics and a M.C.S. in Computer Science, all from the University of Virginia. He is currently a professor in the Department of Mathematical Sciences at Sweet Briar College. In 2006 he published book with Branko Mitrović, entitled *Andrea Palladio: Villa Cornaro in Piombino Dese* (New York, Acanthus Press), which presents a full measured survey of one of Andrea Palladio's most influential works, through 14 fold-out architectural drawings, approx. 16.5 x 23.4 inches each, accompanied by essays which discuss the design of the villa. He also led a three-day research tour of eight of Palladio's villas directly after Nexus '98: Relationships Between Architecture and Mathematics; this workshop was partially supported by a grant from the Graham Foundation for Advanced Studies in the Fine Arts.